Solar Energy and the U.S. Economy

Solar Energy and the U.S. Economy

Christopher J. Pleatsikas,
Edward A. Hudson, and
Richard Goettle IV

Routledge
Taylor & Francis Group
LONDON AND NEW YORK

Acknowledgments

The authors would like to thank the following people for their assistance in researching, writing, and editing this book: Adele Baden, Roger Bezdek, Christopher Dunn, Douglas Funkhouser, Thomais Karidoyanes, John A. Lewis, Malcolm Persen, Patrick Peracreta, and Sherri L. Rose.

First published 1982 by Westview Press

Published 2019 by Routledge
52 Vanderbilt Avenue, New York, NY 10017
2 Park Square, Milton Park, Abingdon, Oxon OX14 4RN

Routledge is an imprint of the Taylor & Francis Group, an informa business

Library of Congress Cataloging in Publication Data
Pleatsikas, Christopher J.
Solar energy and the U.S. economy.
(A Westview replica edition)
1. Solar energy--Economic aspects--United States. 2. Energy policy--Economic aspects--United States. I. Goettle, Richard. II. Hudson, Edward A. III. Title.
HD9681.U62P56 338.5'443'0973 81-23970
AACR2

ISBN 13: 978-0-367-28790-0 (hbk)

ISBN 13: 978-0-367-30336-5 (pbk)

Contents

Tables

1 Introduction

Twice during the 1970s -- the Arab Oil Embargo of 1973-74 and the tight oil market of 1978-79 associated with the political upheaval in Iran -- the U.S. economy experienced severe shocks as a result of massive price increases for imported oil. By 1980 the price for imported crude oil had increased to nearly twenty times the price in 1970 (about $40 per barrel in 1980 as compared to about $2 per barrel in 1970). The increased reliance on foreign sources for oil because of reduced domestic supply and increased domestic demand exacerbated the situation further. The effects on economic performance have been extremely serious. The rapid rise in energy prices was a major cause of the severe recession of 1974-75 and has contributed to sluggish economic performance since 1975. And prices will continue to rise as a result of the elimination of domestic oil price controls in 1981 and the planned phase-out of domestic natural gas price controls by 1985.

For a nation nearly self-sufficient in energy through the mid-1960s and whose economy was based upon cheap and abundant energy resources, the economic implications of rising energy prices cannot be ignored. The nation's reliance on expensive imported petroleum has severely strained our world balance of payments position, compromised our national security, and confronted the nation with difficult choices concerning future patterns of economic growth and energy consumption. Energy analysts may disagree about the magnitude of domestic energy resources and reserves, but there is little likelihood that domestic production of oil and gas (from conventional and unconventional sources) can be increased above current levels for any sustained period. Indeed, it is likely that domestic production of these fuels, which peaked in the 1970s, will continue to decline. Thus, given expectations of future growth in the economy and energy consumption, coupled with the current need for costly petroleum imports, the supply potential and associated economic impacts of alternative non-conventional energy supply technologies must be carefully evaluated.

Several courses of action are open to individuals, corporations, and public entities to adjust to the new conventional energy supply availability and price conditions. These include vigorous expansion of domestic coal, oil, and gas production as well as more rapid development of nuclear fission electric power. However, these courses of action will require an increased commitment of capital to recover resources which are increasingly expensive to exploit and may entail unknown, long-term damage to the environment. Another possible approach to the problem of scarce energy resources involves a commitment to improved energy efficiency and development of renewable energy resources, principally direct and indirect solar sources.

As an expression of commitment to this latter approach, the United States Congress, in October 1974, passed the Solar Energy Research, Development and Demonstration Act. The Act states that:

> The urgency of the nation's critical energy shortages and the need to make clean and renewable energy alternatives commercially viable requires that the nation undertake an intensive research, development, and demonstration program...[1]

Since that time the federal government has spent hundreds of millions of dollars on research, development and demonstration of innovative energy technologies, including solar and other renewable technologies.[2] Also, federal and state governments have implemented numerous financial incentives, such as special investment and individual tax credits, designed to encourage adoption of these technologies. At the present time many innovative technologies are commercially available (e.g., biomass direct combustion,

---

[1]Section 2(a)(9).

[2]In most of the discussions in this study the terms "solar energy technologies" and "renewable energy technologies" are used interchangeably. They refer to solar thermal (active and passive), wind, photovoltaic, ocean thermal (OTEC), hydroelectric, and biomass conversion technologies. This study also follows the practice of many analysts by including geothermal technologies (direct use and electric) as "renewable."

active and passive solar space heating systems). In addition, many other innovative technologies will reach the commercialization stage during the next twenty years. In order to develop long-range energy policies applicable to the 1980s and 1990s, it is imperative that the economic impacts inherent to the use of these innovative energy technologies be fully investigated.

The objective of the study is to assess the long-range national economic impacts of a large-scale commitment to solar technologies and to formulate a set of economically efficient policies toward these technologies.

RESEARCH DESIGN

In designing any U.S. energy policy for the next two decades, the effects of several factors must be taken into account. These include (but are not limited to) environmental considerations, national security, growth in Gross National Product (and other economic variables) and equity (distribution of benefits and costs across regions, racial/ethnic groups, and income groups). This study will investigate the macroeconomic implications over the next twenty years of a large scale solar/renewables energy investment program, including the impact on GNP, prices, and other measures of national economic performance. Thus, this study will provide information on one of the most important factors which needs to be considered in any comprehensive review and/or evaluation of different energy supply policy options.

It has been suggested by some energy analysts that a massive commitment to develop solar and other renewable energy technologies would have negative impacts on the national economy. This negative impact would be a direct result of the high capital costs of solar energy technologies compared to more "conventional" alternatives. These large capital requirements may divert capital from more productive investments. Proponents of solar and renewable energy counter that nuclear and coal-based energy supply technologies are at least as expensive as their solar counterparts, that imported oil drains our nation of needed capital, and that the manufacture and installation of solar technologies will provide much needed jobs for American workers.

To provide a basis for determining the macroeconomic effects of solar energy investments over the 1980 to 2000 time period, several related issues must also be considered:

- How do the macroeconomic implications change as the share of the energy supply market captured by solar and other innovative technologies increases?
- Because considerable uncertainty exists as to the ultimate costs of solar and other innovative technologies, how do the macroeconomic implications change as the prices of these technologies change?
- Because there is considerable uncertainty as to the costs for conventional energy sources over the next two decades, how do the macroeconomic implications of a national commitment to solar energy change as the prices of competing, conventional technologies/fuels change?

To answer these questions, it is necessary to structure the analysis in a multidimensional form, specifying variations in

- conventional energy costs;
- solar market penetration;
- solar technology costs.

This form is reproduced in Table 1-1. This table represents a three dimensional matrix (of dimension 2x2x2). Within the framework of these eight cases or "scenarios," the economic implications of solar market penetration at varied levels of energy supply costs (conventional and solar) are investigated. This study of macroeconomic effects is, thus, formulated as a sensitivity analysis to test the implications of variations in market penetration or energy supply costs in a consistent format. For each of the scenarios, an analysis of several macroeconomic variables is performed:

- trends in real GNP over the 1980-2000 forecast period for each scenario;
- trends in real consumption and investment;
- changes in the structure of economic activity;
- trends in gross productivity;
- relative price effects;
- capital investment trends (including possible impacts on capital markets of solar energy capital requirements).

An important constraint placed upon the research methodology for this study was a resource constraint which limited the number of scenarios that could be analyzed to just eight. In reality, the choice between a "solar energy future" and a "non-solar energy future" involves choices among numerous possible supply mixes

Table 1-1

SCENARIO CONFIGURATIONS

Lower Solar Market Penetration

Conventional Energy Cost Variations	Solar Cost Variations	
	Most Likely Costs	Lower Costs
Baseline Trend for Conventional Costs	1	3
Higher Conventional Costs	2	4

Higher Solar Market Penetration

Conventional Energy Cost Variations	Solar Cost Variations	
	Most Likely Costs	Lower Costs
Baseline Trend for Conventional Costs	5	7
Higher Conventional Costs	6	8

Notes:

1. "Baseline Trend for Conventional Costs" is defined to include moderate real cost increases for electric energy and synthetic fuel capital/non-fuel operating costs and relatively significant real cost increases for all fossil fuels.
2. "Higher Conventional Costs" is defined to include relatively large real cost increases for electric energy and synthetic fuel technologies and very rapid real cost increases for all fossil fuels.
3. "Most Likely Solar Costs" is defined so that moderate real cost decreases for currently available solar and other renewable technologies are achieved by 2000 and most other renewable energy technologies (where feasible) reach cost goals as specified by the U.S. Department of Energy.
4. "Lower Solar Costs" is defined so that costs for solar and other renewable technologies are, in most cases, significantly lower than in the "Most Likely Solar Costs" case.
5. "Lower Solar Market Penetration" is defined as a more moderate increase in solar/renewables market penetration from current levels (to approximately 10 percent of all energy use including as solar all hydro and biomass technoloigies), while "Higher Solar Market Penetration" is defined as a more rapid increase in penetration (to 15 to 17 percent of all energy use).

(e.g., high solar market penetrations could replace or complement large-scale development of nuclear power or synthetic fuels). In short, to determine an optimum energy supply future from an economic standpoint, a large number of configurations of energy supply should be considered and their economic implications evaluated. However, any analysis must simplify the actual set of processes under consideration. The important goal is to simplify in such a way that the core mechanisms remain explicitly included in the analysis. The three dimensions explicitly considered -- solar costs, solar quantities or penetration, and conventional energy costs -- are the core issues. Therefore, an analysis using eight carefully chosen scenarios should still be sufficient to yield major insights into the role for solar energy.

The most effective way to simplify the problem and adequately address the core issues is to design the scenarios around consistent assumptions about the displacement of conventional fuels by solar energy. For example, we assume that in the residential and commercial sectors, solar energy would displace mainly gas and electricity, while in the industrial sector it would displace mainly oil and gas -- see also Chapters 2 and 4 of this study. The displacement assumptions used for this research are not necessarily the only such assumptions that could be used or even necessarily the most probable. However, they have been designed to be realistic and internally consistent. Furthermore, these assumptions allow for direct cross-scenario comparisons of economic impacts.

In summary, then, four points should be emphasized before the precise specification of scenarios is discussed:

- The scenarios which have been designed for this analysis have been specified to reveal consistent economic comparisons. As such, only one of the many possible patterns for the conventional fuel displacement by solar technologies has been evaluated, although there are possible alternative assumptions.
- The scenarios which have been developed for this analysis have been designed for the purposes of conducting a sensitivity analysis. Thus, although they have been derived, in part, from market research, they should not be construed to represent prescriptive definitions of future energy supply mix alternatives. Numerous possible combinations of future energy

supply mixes exist; the scenarios developed for this analysis merely represent selected feasible alternatives.

- The full capital costs for achieving the energy utilization patterns specified in the scenarios have been allocated, to the extent possible, to the appropriate economic sectors;
- Energy utilization in all scenarios is reported in primary energy equivalents.[3] To convert to delivered energy services requires information on link efficiencies for each sector and fuel type.

THE DYNAMIC GENERAL EQUILIBRIUM MODEL

The Dynamic General Equilibrium Model (also referred to as DGEM) will be used to evaluate the economic implications of the scenarios developed for this analysis. It is a model of economic structure and economic growth which can be used to simulate economic reaction to a variety of energy supply assumptions (including costs and sources of supply). As it is a purely national model, no sub-national spatial detail is reported or input. Prices and costs in the model are reported in 1980 dollars, as they are in this report.

The model's sectoral specification is oriented toward energy so that the model provides a framework for the analysis of interdependencies within the energy system as well as of the interdependence between energy and the rest of the economy. In addition, the model provides a comprehensive basis for medium and long term economic and energy forecasting and for the analysis of economic growth and structural change. This model is discussed in more detail in Chapter 2.

OVERVIEW OF THE STUDY

Chapter 2 of this study will present detail about the research design used to analyze energy/economic interactions under different energy supply conditions. This chapter also discusses energy cost assumptions used for the research and the DGEM model structure.

Chapter 3 presents data inputs and outputs for the "Reference" (or benchmark) energy and economic scenario

---

[3]Primary energy equivalent is the volume of energy resources actually extracted, not the value of energy delivered or utilized at the end-use. Solar energy technologies, under this scheme, are valued at the primary energy which they replace.

used for the analysis. This scenario does not necessarily reflect the most likely energy and economic assumptions among the eight scenarios. However, this scenario, because of its design, represents a convenient benchmark for use in measuring changes across the different scenarios.

Chapter 4 discusses the underlying assumptions and implications of the other seven energy/economic scenarios used for this research. In this chapter each of these scenarios is analyzed in detail.

Then, Chapter 5 compares the economics and energy impacts of different energy price and supply assumptions for all eight scenarios, while Chapter 6 discusses the policy implications that can be derived from these comparisons.

2
Study Design and Assumptions

This chapter describes the underlying assumptions used to develop economic/energy scenarios for the analysis of renewable energy. In successive sections we discuss

- the basic methodology for this study;
- the structure and characteristics of the DGEM energy/economic model;
- technology cost assumptions for electric generation, renewable and synthetic fuel equipment;
- Reference Case fuel cost and energy supply assumptions; and
- economic and demographic assumptions used for the Reference Case.

Finally, beginning on page 33, we describe the principal deviations (from the Reference Case) in fuel price, technology costs, and energy supply assumptions used for the additional scenarios developed for this analysis. References for Chapter 2 are included in the Appendix to this book.

METHODOLOGY

The starting point for this analysis is a detailed forecast of the growth and structure of the U.S. energy and economic systems for the period 1980 to 2000. This forecast originated through use of the Dynamic General Equilibrium Model (DGEM), originally developed by Hudson and Jorgenson and subsequently enhanced by Goettle and Hudson. DGEM permits simulation of future energy and economic activity under a comprehensive set of assumptions covering changes in demographic, economic, or energy conditions. This initial projection, the Reference Case, is described in detail in Chapter 3.

Among its many assumptions, the Reference Case incorporates:

- fuel price forecasts for domestic and imported energy supplies;
- costs for selected electric generation, synthetic fuel, and end-use technologies;
- production patterns for domestic oil and gas; and
- market penetration rates for specific energy supply and conversion technologies.

For this analysis, certain of these assumptions are varied systematically so as to examine the effects of such changes on the economic and energy systems. Since, in each scenario (or case), a specific set of energy conditions is altered, comparison of economic performance between scenarios provides an estimate of the economic effects of these energy changes. In this way, cause and effect information is generated -- the cause being the energy change, the effect being the change in economic performance. The economic effects, direct and indirect, immediate and delayed, are captured by this procedure.

The focus of the analysis is the effects of changes in renewable energy costs and use conditions. The two variables of direct concern are, therefore, the unit cost of energy obtained from renewable sources and the quantity of energy obtained from renewable sources (referred to as the market penetration by renewables). However, since the economics of renewables are so closely linked to the cost of conventional energy (because renewables will displace energy from conventional sources) and since the future cost of conventional energy is uncertain, a third major variable, the cost of conventional, nonrenewable energy, is included in the analysis.[1] This gives the following three factors or variables as the basis for the analytical design:

- unit cost of energy services from renewable sources;
- quantity of energy services obtained from renewable sources;
- unit cost of energy services from nonrenewable sources.

Each of these variables is specified as being at one of two possible values. Renewables costs can be at the "Most Likely" level or at a different, lower level.

---

[1]Nonrenewable energy is the energy obtained from fossil and nuclear fuels.

Renewables market penetration can be at the "Lower" level or at a different, higher level. Nonrenewables costs can be defined by the "Baseline Trend" or by a different trend which results in higher costs over the forecast period. These variations are intended to incorporate the principal uncertainties concerning these energy variables. The Reference Case combines assumptions about all three variables (renewables costs, renewables penetration, nonrenewbles costs) which represent their respective most realistic estimate, or most likely outcome. Relative to this case, the most likely variation is judged to be down, for renewables costs; higher, for renewables penetration; and higher, for prices of conventional energy. Therefore, for each energy variable, the two outcomes considered in this study represent what is judged to be most likely value together with the most likely direction for deviation from this value.

In combination, there are eight possible configurations of the energy system (or at least that portion of the system covered by these causal variables). These eight possibilities are explicitly examined. The result is eight forecasts for the economy, each forecast based upon one possible set of energy conditions.

These different energy specifications, each with a corresponding economic forecast associated with it, are represented systematically in Tables 2-1 and 2-2. The first Table sets out the different combinations of conditions in tabular form. For the Lower market penetration level for renewables, there are four scenarios, corresponding to all possible combinations of renewable and nonrenewable energy costs. Similarly, there are four possible energy cost scenarios for higher levels of penetration by renewables. Table 2-2 shows the same information in tree form with the eight possible situations being constructed in terms of the three component parts. Both tables show the notation or labelling used in the remainder of this study to refer to each scenario. Thus, R is the Reference Case while HC is the scenario in which nonrenewables costs are high (.i.e., High Cost) but other variables are at their Reference Case levels. HPLHC is the case with high renewables penetration, low renewables costs and high nonrenewables costs. The other cases are labelled analogously.

STRUCTURE AND CHARACTERISTICS OF DGEM

The Dynamic General Equilibrium Model (DGEM) is a simulation model of the structure and growth of the U.S. economy. It combines a model of macroeconomic growth with a multisector, input-output model based on flexible

Table 2-1

THE EIGHT ENERGY SCENARIOS DEVELOPED FOR THIS STUDY

Lower Market Penetration of Renewables

Fuel Prices and Technology Costs for Nonrenewable Energy	Fuel Prices and Technology Costs for Renewable Energy	
	Most Likely	Lower
Baseline Trend	R	LC
High Trend	HC	LHC

Higher Market Penetration of Renewables

Fuel Prices and Technology Costs for Nonrenewable Energy	Fuel Prices and Technology Costs for Renewable Energy	
	Most Likely	Lower
Baseline Trend	HP	HPLC
High Trend	HPHC	HPLHC

Note: The abbreviation appearing for each entry is used throughout the analysis to identify the particular scenario. The abbreviations are defined as follows:

R	=	Reference Case
LC	=	Lower Costs for Renewables Case
HC	=	Higher Costs for Nonrenewables Case
LHC	=	Lower Renewables Costs, Higher Nonrenewables Costs Case
HP	=	Higher Renewables Penetration Case
HPLC	=	Higher Penetration and Lower Costs for Renewables Case
HPHC	=	Higher Renewables Penetration, High Nonrenewables Costs Case
HHPLHC	=	Higher Penetration and Lower Costs for Renewables and High Costs for Nonrenewables Case

Table 2-2

DIAGRAM OF EIGHT COMBINATIONS OF ENERGY CONDITIONS

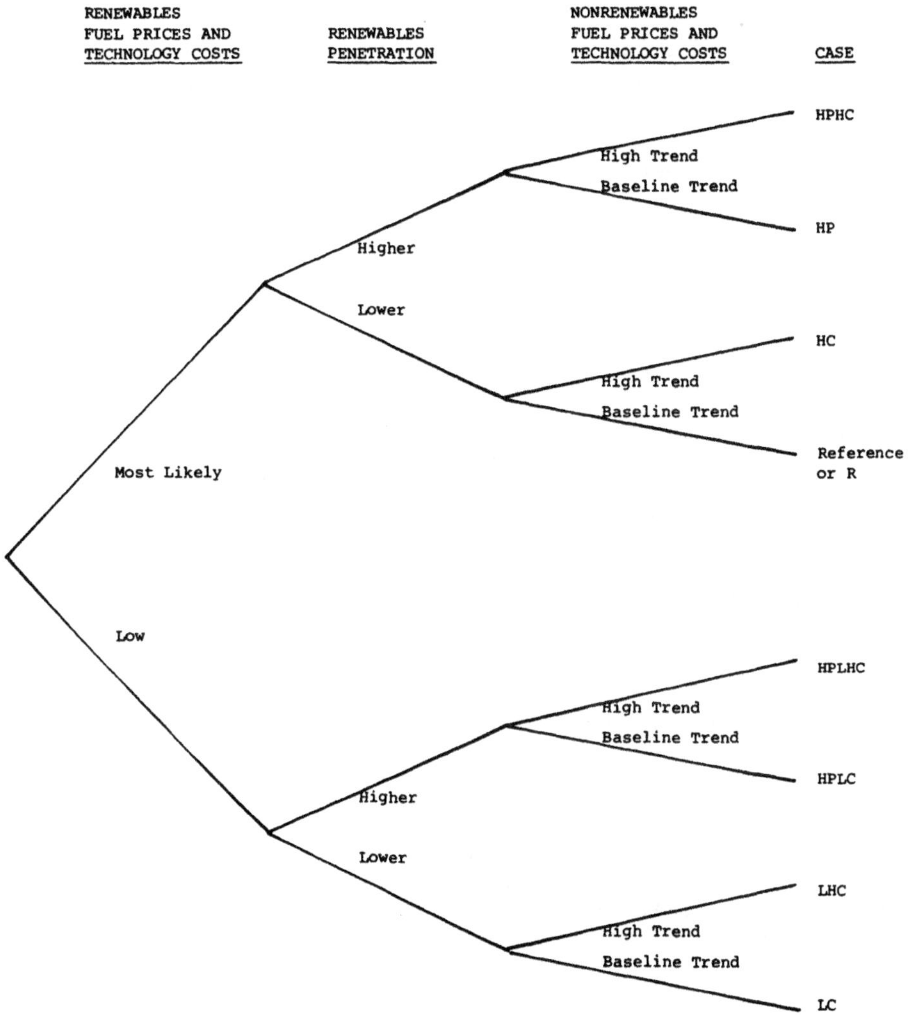

coefficients. For each year, it models economic activity on a sectoral basis and integrates these into a consistent whole. There are nine producing sectors (see Table 2-3) included in the model. Five are energy related and the remainder are non-energy producers. New technology energy supply, conversion, and end-use activities are considered explicitly in the model. They are divided into five categories:

- coal supply and demand substitutes;
- petroleum supply and demand substitutes;
- electricity supply substitutes;
- electricity demand substitutes;
- gas supply and demand substitutes.

New energy sources first are specified in detail and, then, are aggregated into these five categories see Table 2-4).

Inputs into production and final demand purchases are provided by the nine main producing sectors. In addition, there are three other sources of inputs to the economy (capital, labor, and competitive imports) and four categories of final demand for goods and services (personal consumption expenditures, investment, government purchases, and exports). These activities are organized into a matrix of interindustry transactions.

DGEM forecasts present a dynamic-equilibrium view of the U.S. economy. Balance or consistency is required to hold in several senses. First, price formation must be such that sectoral output prices cover average production costs, including a normal rate of return. Second, sectoral output quantities must match exactly the quantity of that good or service required for input into production together with the amount used to satisfy final demand. Third, prices and quantities must be such that the revenue received by a sector is exhausted totally by payments to inputs, including income to capital, and payments to governments. Fourth, the demands for labor and capital must be consistent with the supplies of these resources, the former being determined within the context of household behavior (i.e., labor-leisure choice) and the latter being determined through a dynamic saving-investment mechanism.

Activity patterns within DGEM are represented by econometric submodels. Also, the production requirements for new energy sources are integrated into the overall structure through use of technological (i.e., fixed-coefficient) submodels. The econometric submodels for each of the conventional producing sectors incorporate the patterns of behavioral and technical responses observed for these activities. Collectively, these submodels provide the framework for analysis of output

Table 2-3

INTERINDUSTRY TRANSACTIONS SECTORS
IN THE DYNAMIC GENERAL EQUILIBRIUM MODEL

- Coal mining
- Petroleum and natural gas extraction
- Crude oil refining
- Gas utilities
- Electric utilities
- Construction, agriculture, and non-fuels mining (including uranium mining)
- Manufacturing
- Transportation
- Services

Table 2-4

TECHNOLOGICAL DETAILS OF ENERGY PRODUCTION AND CONSUMPTION WITHIN DGEM

Displacement of:	By:
Conventional Coal Supply and Demand	Biomass Resources Direct Biomass Use by Industry Geothermal Process Heat
Conventional Petroleum Supply and Demand	Oil from Shale Coal Liquids Biomass Liquids Solar/Oil Space Heat Solar/Oil Process Heat Geothermal Process Heat Direct Biomass Use by Households Direct Biomass Use by Industry
Conventional Central Station Electricity Supply	Advanced Coal Systems Low-head Hydroelectric Nuclear Geothermal Electric Biomass Electric Solar Thermal Electric Centralized Wind Centralized Photovoltaics
Conventional Electricity Demand	Decentralized Wind Decentralized Photovoltaics Solar/Electric Heating and Cooling Solar/Electric Water Heat
Conventional Gas Supply and Demand	Medium Btu Gas from Biomass Medium Btu Gas from Coal High Btu Gas from Coal Solar/Gas Water Heat Solar/Gas Space Heat Solar/Gas Process Heat Direct Biomass Use by Households

price formation and, with the aforementioned balance conditions, determine the system of relative prices within the economy. The submodels also determine the particular pattern of input purchases that minimize production costs given prevailing prices. Thus, input-output coefficients and sectoral productivity change are endogenous, being functions of relative prices and price-induced technological change. Similarly, the pattern of consumption spending is modeled in a flexible manner so as to incorporate price-induced flexibility in purchase patterns. These features permit the incorporation of key aspects of producer and consumer behavior, such as substitution or complementary relationships between inputs, and adjustment in the pattern of consumer expenditures.

As a result, some of the principal adjustment mechanisms in the economic system are explicitly incorporated in the model. For energy, these features incorporate the analysis of the patterns of use of energy as an input to each type of production as well as the patterns of energy use in consumption and other final demand activities. For the economy in general, this specification provides for the incorporation of a full supply representation of production, including input requirements, input constraints, input adjustment possiblities and productivity. Further, the general equilibrium specification ensures that the direct and indirect effects throughout the economy of energy (or other) changes are accommodated.

Simulations from DGEM indicate the patterns of expenditures that are consistent with what is achievable from production, i.e., the supply possibilities of the economy. Therefore, output is dependent on input availabilities, sectoral rates of productivity growth and production requirements, and their combined implications for overall economic productivity. In each period, input prices in conjunction with the supply of inputs determine incomes, which primarily are composed of incomes to capital and labor. Incomes together with prices, wealth, rates of return, and other influences, determine expenditures. Final spending and interindustry input patterns, also reflecting relative prices, determine total industry outputs and, so, demands for inputs. If demand and supply quantities are not equal in both product <u>and</u> factor markets, then prices and rates of return adjust so as to achieve this equality. Over time, the saving/investment mechanisms influence the productive potential of the economy through their effects on wealth and capital accumulation. The balance between savings and investment determines the rate of return and the rate of growth of the capital stock. Thus, investment in each period establishes the availability of capital in the following period.

The resulting forecast includes the following information:

- final demand expenditures for each type of good or service that reflects income, prices, and other determinants;
- levels of real economic activity that are feasible in terms of the supply and demand characteristics of each sector and of the economy;
- a relative price structure that is consistent with demand, supply, and resource configurations;
- a supply structure that reflects both patterns of inputs and productivity change in each sector and, through time, the availability of these inputs and resources to the economy;
- economic growth as a sequence of one-period equilibrium giving demand, supply, and the relative price for each commodity and input.

Treatment of Capital Investment in DGEM

This study is concerned mainly with the impact of the capital requirements for solar and conventional energy supplies on U.S. economic performance. In particular, it is the change in capital investment required to achieve a specified level of solar market penetration that is of interest. Therefore, it is useful to discuss the treatment of capital investment in the Dynamic General Equilibrium Model.

The input of capital services to the energy sectors is directly related to the outputs specified for those sectors. For example, by exogenously specifying a level of output for the energy extraction sectors in the model (i.e., coal, oil, gas), the model will respond by diverting to those sectors the volume of capital services required to sustain the stipulated output. Energy price levels also affect capital and other requirements. In particular, a rising real price for energy is an indication that greater factor inputs are required to produce each additional unit of energy resources. For these reasons, the exogenous specification of energy supply and prices that will be used for runs of the DGEM system replicates the exploitation of increasing marginal cost energy resources.[2]

---

[2]Defining capital services inputs to the oil, gas and coal sectors on the basis of output and resource price represents an alternative to simulating the effect of a resource extraction curve.

The DGEM model does not distinguish between energy using capital services and other "productive" capital services for the residential, commercial, transportation and industrial sectors. In this manner, the demand for capital services in these sectors is more a function of output and input prices than of energy use (producers -- both household and otherwise -- are free to utilize the mix of capital services which may best achieve output objectives). In this sense, the model is more flexible and realistic than if energy use in all energy consuming sectors were tied in fixed relationships to output, since, in the long run, labor and/or capital may be substituted for energy.

In order to determine the economic impact of the capital requirements for solar/renewable energy sources (or any other supply mix), deviations from the value of depreciated capital stocks from some base value must be developed. This requires that capital investment information be input to the model in an appropriate format. The form in which capital requirements are specified for the scenarios is as follows:

- For energy extraction sectors capital requirements are determined endogenously in the model using domestic output and price information introduced from outside the model.
- For the gas utility sector capital requirements are specified endogenously based on capital flow trends. Capital services input to this sector do not differ significantly across the different scenarios.
- For the electric utility sector, a submodel may be used to specify investment requirements for conventional technologies. These requirements can be generated using traditional capital/output relationships ($/kW). Similar information is generated exogenously for non-conventional electric technologies. Transmission and distribution investment are calculated within the model.
- For the petroleum refining sector, the model determines capital service inputs endogenously based on the assumption that 50 percent of all imported oil and all of the domestically-produced oil is refined in the U.S. Thus, refining capacity will implicitly vary based on domestic output and demand requirements.
- For all new technology energy resource sectors, the effect of solar and other capital investment must be input to the model in the form of pre-specified coefficient and cost relationships.

- For energy end-use investments, DGEM calculates capital service requirements based on energy demand by each sector and the functional relationships specified in the model among input prices and factor demands.

TECHNOLOGY COST ASSUMPTIONS

Given the structure of DGEM, cost information for certain energy conversion technologies is required as an input for the construction of each energy-economic scenario. These technologies include

- renewables;
- electric generation;
- synthetic fuels.

Cost information was developed based on a comprehensive review of published and unpublished documents from public and private sources. Consistent with the research design for this study, two cost scenarios were specified for each technology category. For renewables these cost scenarios are:

- Most Likely (or Baseline) Costs Scenario, representing "most probable" costs over the 1980-2000 projection period. This scenario presumes that, for nearly all renewable technologies, significant cost reductions are achieved over current technology costs. In many cases federal cost goals were used to represent ultimate technology costs for this case.
- Low (Optimistic) Cost Scenario, representing lower cost levels which may still be achievable in future years if certain technological or production advances (beyond those specified for the "Most Likely" Case) are realized.

For electric technologies, the two cost scenarios are:

- Baseline Trend Cost Scenario, representing a continuation of past trends (since the late 1960s) with an attenuation of recent past real inflation. This attenuation represents an implicit assumption that enviroinmental standards will not be tightened further and that cost increases caused by construction delays or lengthened construction schedules will get no worse.
- High Cost Trend Scenario, representing a continuation of recent past trends with little

or no attenuation of real inflation. This scenario represents an implicit assumption that environmental standards for these technologies will be further tightened and that further cost increases will occur as a result of construction delays and lengthening of construction schedules.

The problem in developing cost estimates for synthetic fuel/shale oil facilities is that no commercial scale facilities have yet been built. Thus, all cost estimates are based on engineering studies which attempt to estimate costs by scaling up pilot or demonstration facilities. As research on various synthetic fuel processes has proceeded, additional investment requirements (especially to correct previously unforeseen environmental or process-related problems) have been identified. Thus, projected cost estimates for synthetic fuel facilities have escalated at a particularly sharp rate over the past decade. Because there remain unresolved environmental and socioeconomic[3] problems with synfuel/oil shale technologies, it is likely that costs for these facilities will continue to increase at a rate faster than that for prices in general in the economy. For the lower bound (Baseline Trend), the most recent (1980-1981) cost estimates for these facilities have been used in conjuction with a 1.5 percent real inflation rate for 1980-1990 and a 1 percent real inflation rate for 1990-2000. For the upper bound (Higher Cost Trend scenarios), rapid real escalation in costs was assumed -- at a rate of 3 percent per year (even higher than the High Cost Case real cost increases for coal-fired electric utility plants).

A summary of costs developed for this study is included as Table 2-5. Some important aspects of the cost estimates for these technologies are:

- All estimates are national average estimates.
- Where large discrepancies among sources existed, greater reliance was placed on more recent estimates.
- The focus of this research was on technologies that are currently commercially available or could possibly make a significant contribution to energy supply before the year 2000.

---

[3]For example, many of these plants would be built in sparsely populated areas of the Western United States, where premium wages have to be offered to construction/operations workers to induce them to relocate.

Table 2-5A

SUMMARY OF CAPITAL COST ESTIMATES FOR SELECTED RENEWABLE TECHNOLOGIES (1980$)

Technology	Units	1980 Cost	Year 2000 Costs	
			Most Likely	Low (Optimistic)
Solar Active Residential	$\$/10^6$ Btu	250	250	175
Solar Active Commercial	$\$/10^6$ Btu	225	225	125
Passive Solar Residential	$\$/10^6$ Btu	200	160	140
Passive Solar Commercial	$\$/10^6$ Btu	180	180	100
Small WECS[a]	$/kW	3,000	1,500	1,200
Large WECS	$/kW	2,500	1,000	800
Solar AIPH[b]	$\$/10^6$ Btu	200-250	80-90	65-70
Solar Thermal Electric				
with storage	$/kW	6,400[e]	1,850	1,600
without storage	$/kW	4,400[e]	1,600	1,200
OTEC[c]	$/kW	4,300[f]	3,660	2,100
PV[d] - Residential (no storage)	$/kW	10-12,000	1,450	1,100
PV - Utility (no storage)	$/kW	10-12,000	1,450	1,100
Geothermal (Electric)	$/kW	650	1,587	1,587
Large Hydroelectric				
Pumped Storage	$/kW	450	550	550
Conventional	$/kW	870	1,065	1,065
Small Hydroelectric	$/kW	1,500	1,500	1,500
Biomass				
Electric	$/kW	1,500	1,500	1,200
Steam	$\$/10^6$ Btu	8.00	8.00	7.25
Gasification	$\$/10^6$ Btu	30-36	30-36	27-32

[a]WECS = Wind Electric Conversion Systems
[b]AIPH = Agricultural/Industrial Process Heat
[c]OTEC = Ocean Thermal Electric Conversion
[d]PV = Photovoltaic
[e]1985 Cost.
[f]1990 Cost.

Table 2-5B

SUMMARY OF CAPITAL COST ESTIMATES FOR SELECTED ELECTRIC AND SYNTHETIC FUEL NONRENEWABLE ENERGY TECHNOLOGIES (1980$)

Technology	Units	1980 Cost	Year 2000 Costs	
			Baseline Trend	High Trend
Coal w/FGD[a] (Electric)	$/kW	825	1,005	1,352
Oil w/FGD (Electric)	$/kW	700	853	1,146
Nuclear LWR[b] (Electric)	$/kW	990	1,269	1,783
Gas Turbine Peaking (Electric)	$/kW	235	235	315
Oil/Gas Combined Cycle (Electric)	$/kW	440	537	540
Coal Gasification CC[c] (Electric)	$/kW	825	1,005	1,350
Coal AFB[d] (electric)	$/kW	NA	885	1,190
Nuclear LMFBR[e] (Electric)	$/kW	NA	1,688	2,110
Coal - Low Btu Gasification	$/10^6 Btu	11.00	14.11	19.84
- Medium Btu Gasification	$/10^6 Btu	12.50	16.03	22.58
- High Btu Gasification	$/10^6 Btu	24.00	30.49	43.34
- Liquifaction	$/10^6 Btu	16.00	20.52	36.07
Shale Oil	$/10^6 Btu	15.00	19.22	30.67

[a]FGD = Flue Gas Desulfurization.
[b]LWR = Light Water Reactor.
[c]CC = Combined Cycle.
[d]AFB = Atmospheric Fluidized Bed.
[e]LMFBR = Liquid Metal Fast Breeder Reactor.

- For non-electric technologies, capital costs are derived by dividing total capital investment by annual Btu output (in an average year of operation).
- For electric technologies, all capital costs are expressed in terms of $/kW (kW stated in terms of rated capacity).
- All dollars have been converted to 1980 dollars.
- Total capital investment is defined to be inclusive of land costs, architectural and engineering fees, contingency costs and real AFDC (allowance for funds used during construction) costs.
- All capital costs represent costs for a facility completed in the year for which the estimate was made.

All nonrenewable energy end-use equipment costs (e.g., furnaces, boilers) were defined on the basis of information from the Brookhaven National Laboratory.[4]

FUEL COST AND ENERGY SUPPLY ASSUMPTIONS IN THE REFERENCE CASE

Two nonrenewable fuel price trends have been developed for this research. They are termed the "Baseline Trend" and the "High Trend" (see also Table 2-1). Both trends were based on total decontrol of domestic oil prices in early 1981 and partial decontrol of domestic natural gas wellhead prices by 1985. These two nonrenewable fuel price trends correspond to two nonrenewable energy technology cost trends also developed for this research. These are also termed "Baseline Trend" and "High Trend" (see also Table 2-5). For this research the "Baseline Trend" for nonrenewable fuel price was used only in conjunction with the Baseline Trend nonrenewable energy technology cost assumptions. Similarly, the High Trend for nonrenewble fuel prices was only used in conjunction with the High Trend for nonrenewable energy technologies. There is nothing deterministic in these pairings. They were chosen merely to represent the range of conventional (or nonrenewable) energy costs that may occur over the next two decades.

For the Reference Case, the Baseline Trend is used for nonrenwable energy costs. Technology Costs for this case are noted in Table 2-5. The Baseline Trend

---

[4]Brookhaven National Laboratories, Evaluation of Technological Data in the DFI and PIES Models, 1979.

nonrenewable energy price assumptions used for this case are presented in Table 2-6. The world oil price is assumed to increase from $34 per barrel (1980$) in 1980 to almost $64 per barrel (1980$) by the year 2000. For the United States, growth in average crude oil costs is highest over the period 1980 to 1985, averaging 7.4 percent annually; from 1985 to 2000 oil costs grow at an average annual rate of 3.1 percent. Full decontrol of domestic oil prices occurs in 1981 and the decontrol assumption includes the provisions of the Crude Oil Windfall Profits Tax of 1978. Partial decontrol of natural gas prices occurs in 1985 in accordance with the Natural Gas Policy Act. Growth in wellhead gas prices averages 8.0 percent annually through 1984 and experiences a sharp increase in 1985 as many categories of natural gas become fully deregulated. After 1985, gas prices grow at slightly faster rates than do oil prices but do not reach "Btu-parity," even by the year 2000. Coal price growth is moderate over the entire projection horizon. This assumption is predicated on an increasing share of total production from western and nonunion producers, economies of scale and productivity improvements in future production, and moderations of both wage and environmental control cost increases. Growth in the price of nuclear fuel reflects the adequacy of relatively low-cost uranium supplies given the moderate growth of nuclear generation capacity that is projected for the future. Biomass resource prices increase more rapidly than coal prices, reflecting the supply and demand pressures that are likely to exist as fuel uses for these commodities become more prevalent.

Domestic oil and gas production levels for the Reference Case are summarized in Table 2-7. This information was obtained from "Analysis of Current Trends in Petroleum and Natural Gas Production," a report by the Comptroller General, General Accounting Office (GAO), 7 December 1979. Slight revisions were made to the GAO gas estimates for the 1986-1990 period to reflect more favorable gas supply conditions following decontrol. In spite of the time since their publication, these estimates are still representative of "middle-range" estimates for future production possibilities. For example, for conventional liquids production, they are neither as pessimistic as those reported by the Office of Technology Assessment[5] nor as optimistic as those

---

[5]Office of Technology Assessment (OTA), U.S. Congress. "World Petroleum Availability: 1980-2000." October 1980. Production estimates are in the range of 5.3 to 7.6 million barrels per day (MMBD) for 1990 and 4.0 to 7.0 MMBD for 2000.

Table 2-6

REFERENCE CASE FUEL PRICES
(1980 Dollars per Million Btu)

Fuel and Source	1980	1985	1990	1995	2000
Crude Oil Refiner Acquisition Cost:					
Domestic	4.14	6.90	8.16	9.57	10.98
Imported	5.86	6.90	8.16	9.57	10.98
Composite	4.83	6.90	8.16	9.57	10.98
Natural Gas, Wellhead	1.49	3.62	4.37	5.24	6.14
Coal, Minemouth	1.12	1.24	1.38	1.53	1.70
Biomass Resources	2.29	2.72	3.24	3.85	4.58
Nuclear Fuel Input	0.75	0.82	0.93	1.04	1.15

Table 2-7

ANNUAL U.S. OIL AND GAS PRODUCTION ASSUMPTIONS

Year	Petroleum and Natural Gas Liquids (MMBD)	Natural Gas (Trillion Cubic Feet)
1980	10.2	19.0
1985	8.9	17.1
1990	8.0	17.9
1995	8.0	16.8
2000	8.5	16.6

Notes:

Liquids production includes Lower-48 and Alaska, onshore and offshore, and enhanced recovery.

Gas production includes Lower-48 and Alaska.

MMBD = Million Barrels per day

developed by the Energy Information Administration.[6] The GAO estimates do appear somewhat more optimistic regarding future gas production than those reported by many other private and public agencies. However, they are well within the estimated production range projected by the American Gas Association, and they seem achievable given progressive decontrol of natural gas prices.

The final set of energy assumptions relates to the levels of utilization of selected energy technologies. Table 2-8 shows technology-specific projections for electric generation. In the electric sector, three technologies -- new coal facilities, nuclear, and geothermal -- account for most of the future capacity growth as electricity demand growth slows and oil- and gas-fired capacity is retired. Nuclear capacity is estimated to expand from 55.2 gigawatts electric (GWe) in 1980 to 160.6 GWe by the year 2000.[7] Geothermal capacity grows much more rapidly, from 0.7 GWe to 10.0 GWe, over the same time horizon. Over the forecast period, coal-based generation is assumed to carry the full burden of adjustment in meeting electricity demand (which is determined within the model). Thus, all other components of the electric generation mix, centralized and decentralized alike, are fixed by assumption. New coal electric generation technologies are introduced over the forecast period, and cost and conversion efficiency inputs to the scenarios reflect these introductions.

By the end of the century, the installed capacities of biomass electric and low-head hydropower for the Reference Case are projected to be 2.8 and 6.0 GWe, respectively. Solar thermal, wind and photovoltaic central stations account for the remaining capacity. In 2000, the rated capacities of these technologies are estimated to be 1.6, 9.5 and 3.6 GWe, respectively. These correspond to 16 solar thermal facilities, over 150 "wind farms," and more than 360 photovoltaic power plants. Thus, the forecast incorporates an increasing commitment toward the use of renewables technologies in electricity supply, in spite of their early cost disadvantages.

---

[6]Energy Information Administration (EIA), U.S. Department of Energy. 1980 Annual Report to Congress, Volume Three. March 1981. Under the mid-price conditions, conventional liquids production is estimated to be 9.6 MMBD in 1990 and approximately 10.0 in 2000.

[7]This increase represents capacity currently under construction or on order as of mid-1981.

Table 2-8

ELECTRIC INPUTS BY ENERGY SOURCE
(Quadrillion Btu)

Central Station Inputs	1980	1990	2000
Biomass	0.010	0.058	0.131
Oil	3.170	1.650	0.900
Natural Gas	3.812	2.186	1.312
Nuclear	2.727	6.560	8.330
Hydroelectric:			
Conventional/ Pumped Storage	3.101	3.349	3.703
Low-head	0.001	0.100	0.412
Geothermal	0.108	0.624	1.608
Solar Thermal	--	0.003	0.050
Wind	--	0.010	0.300
Photovoltaics	--	0.005	0.080

Commercial production of synthetic liquids and gases is not expected to begin until the middle 1980s. Also, for a variety of technical, economic, and regulatory reasons, production increases are not expected to achieve the 1992 national objective of 2.0 million barrels daily (MMBD). However, it is assumed that oils from shale and coal liquids will contribute over 1.1 MMBD by the year 2000. Alcohol fuels yield an additional 162,000 barrels daily. Synthetic liquids thus provide 13.3 percent of total domestic liquids supply by the end of the century. Synthetic gas production is not expected to achieve the commercial successes of the liquids technologies, given the high cost estimates for synthetic gas and the availability and price of gas from conventional sources. However some commercial development is assumed for both decentralized, medium-Btu gasification (from coal and biomass) and the production of pipeline-quality gas from coal. Measured as oil equivalent, it is assumed that these technologies provide approximately 313,000 barrels per day in the year 2000 (see Table 2-9).

Use of solar and other renewables technologies is projected to increase substantially over the remainder of the century (see Table 2-10). Delivered energy from these sources rises from 1.8 quadrillion Btu (quads) in 1980 to 4.1 quads by the year 2000. Wood, wood-waste products, and other biomass resources provide a large portion of the total dispersed renewables supplies. It is estimated that wood use in the residential sector satisfied the space-heating requirements for the equivalent of more than two million households in 1980. This should increase to four million households by 2000. Also, industrial use of biomass is expected to increase, although at a relatively slow rate.

In the Reference Case we estimate that residential and commercial use of active solar technologies will grow at an average annual rate of 24.3 percent through the end of the century. Water heat applications will remain the dominant use in these sectors. However, solar space conditioning (active and passive) is expected to increase from approximately a twenty percent share of total solar use in 1980 to forty percent in the residential/commercial market by the year 2000. In 1980, an estimated one-half million homes were equipped with solar thermal technologies. By 2000, this number is projected to increase to almost 32 million, or 28.6 percent of the total number of households. The commercial sector will account for an additional 13.6 million units of equivalent installed solar capacity. In the agricultural and industrial sectors, active solar is projected to grow at approximately the same rate as commercial applications, reaching 0.2 quads by the year

Table 2-9

SYNTHETIC FUEL PRODUCTION LEVELS
(Quadrillion Btu of Output)

	1985	1990	1995	2000
Synthetic Liquids				
From Coal	0.106	0.635	1.059	1.694
From Biomass	0.027	0.066	0.104	0.189
Shale	0.085	0.276	0.442	0.708
Total	0.218	0.977	1.605	2.591
Synthetic Gases				
Medium Btu from Coal	0.018	0.036	0.072	0.108
Medium Btu from Biomass	0.018	0.035	0.070	0.105
High Btu from Coal	--	0.150	0.225	0.450
Total	0.036	0.221	0.367	0.663
TOTAL SYNTHETIC FUELS	0.254	1.198	1.972	3.254

Conversion Efficiencies:		
	Coal Liquids	0.63
	Biomass Liquids	0.70
	Medium Btu Gas - Coal	0.75
	Biomass	0.90
	High Btu Coal Gas	0.65

Table 2-10

DISPERSED RENEWABLES[a]
(Quadrillion Btu of Delivered Energy)

	1980	1990	2000
Solar Thermal			
Residential/Commercial	0.009	0.131	0.700
Industrial	--	0.032	0.200
Biomass Combustion			
Residential	0.200	0.280	0.360
Industrial	1.600	1.800	2.800
Industrial Geothermal	--	0.004	0.010
Decentralized Wind	--	0.002	0.017
Decentralized Photovoltaics	--	0.001	0.007
TOTAL	1.809	2.257	4.094

[a]Converted to primary energy equivalent at a thermal efficiency equal to 1.0.

2000. Geothermal use by industry represents only a small source of total renewables supplies, with location rather than cost being a more important consideration for future penetration. Finally, commercial success is envisioned for both decentralized wind and photovoltaic systems. In quads, their combined contribution appears to be quite small. However, by 2000, total installed capacity for these technologies exceeds 3 GWe. This corresponds to approximately 190,000 ten kW (kilowatt) wind systems and 234,000 five kW photovoltaic systems for an ownership rate equivalent to almost 0.4 percent of all households.

DEMOGRAPHIC AND ECONOMIC ASSUMPTIONS IN THE REFERENCE CASE

In developing a "Reference Case" forecast, DGEM requires input information on future demographic trends, rates of unemployment and domestic (overall) price inflation, government expenditure and revenue policies, and export patterns. For alternatives to the Reference Case, inflation and government revenue positions vary endogenously while adjustments in government purchases and/or deficits and the economy's export patterns are specified as part of the analysis. Generally, only population and the unemployment rate remain fixed at their Reference Case values.

The Census Bureau's Series II population projections form the basis for the demographic trends incorporated into this projection. These assumptions are summarized in Table 2-11. The rate of population growth under the Series II assumptions shows a steady decline toward, although not reaching, a zero rate. Of importance is the changing age structure of the population. Growth in the working-age population (ages 16 to 65) slows dramatically through the middle nineties. This has significant implications for growth in the civilian labor force and, so, for the overall rate of productivity advance (both of which are endogenous to all the scenarios developed for this study). From 1972 through 1980, growth in the working-age population was almost double that of the population as a whole -- 1.53 percent per annum versus 0.80 percent. In combination with significant changes in relative prices for certain factor inputs (e.g., energy and materials), this growth in the working-age population permitted a rapid expansion in the civilian labor force without a commensurate increase in real compensation. In part, this trend contributed to the observed deterioration in the rate of growth in productivity. The productivity effects of this expansion were exacerbated because employment growth was strongest in the least productive industries (e.g., government and services) within the

Table 2-11

ECONOMIC/DEMOGRAPHIC ASSUMPTIONS

Category and Variable	1980	1990	2000
Demographic (in millions)			
Population	222.5	243.5	260.4
Population, 16-65	143.4	155.3	167.5
Households	80.8	95.5	110.8
Government Purchases[a] (in Billions of 1980$)			
Federal Defense	126.0	191.3	263.3
Federal Non-Defense	66.4	72.6	80.2
Federal Total	192.4	263.9	343.5
State/Local	322.4	383.5	423.6
Total Government	514.8	647.4	767.1
Export Trends (in Billions of 1980$)			
Real Exports	286.7	393.2	557.2

[a]National Income and Product Accounts Definitions used.

	1980-82	1982-85	1985-90	1990-2000
Unemployment Rate (in percent) Average	7.0	7.0	6.7	6.2
GNP Price Deflator (in percent) Growth per year	9.25	8.5	7.3	6.4

economy. With the Series II projection, slower growth in the working-age population suggests that such a pattern is unlikely to recur with similar intensity. Finally, it is projected that household size will continue to decline, reflecting trends toward smaller families and single-person households.

The rate of unemployment is assumed to follow a cyclical pattern along a steadily declining long-run trend. The long-run reduction in the unemployment rate corresponds to the tightening of labor markets that would occur naturally as a result of slower growth in the working-age population. The cyclical pattern around such a trend is based on an examination of recent empirical evidence.

The inflationary pressures within the economy are assumed to abate gradually over the remainder of the century. The process of reducing inflation begins slowly and becomes progresively more effective through time. In part, this is due to the longer-term (post-1985) moderation in energy price increases assumed for the Reference Case forecast. In addition, the economic consequences of slower growth in the working-age population are likely to encourage improvements in productivity growth. These, in turn, favorably affect unit production costs and, so, aid in reducing inflation. Finally, change in government fiscal, monetary, and regulatory policy is directed, with increasing success, toward ameliorating inflation and its consequences. A summary of the unemployment and inflation projections also appears in Table 2-11.

Government transfers and tax revenues are projected to rise approximately in line with the economy as a whole. This, coupled with slower growth in government expenditures, alters the trend in budget deficits so that a smaller portion of total private saving need be absorbed by government. Private investment and the process of capital formation are favored by this change alone. However, there is also a significant change projected for the structure of taxation. For the most part, changes in taxes on labor income are estimated to be neutral. Increases in certain federal, state, and local taxes on labor income are compensated by reductions in federal personal income tax rates. The tax reductions that do occur are directed toward stimulating increased savings and investment by businesses and households. The combination of structural changes in government transfer programs, slower growth in government purchase, and new tax incentives favoring capital inputs are intended to encourage investment at the expense of current consumption and improve productivity growth.

The final economic assumption concerns the future growth of real exports. In each decade since 1950, U.S. export growth has accelerated; the export share of real GNP has increased from 4.4 to 10.9 percent during these thirty years. Over the fifties and sixties, U.S. producers benefitted from increasing incomes abroad, a relatively stable currency, increasing diversity and innovation in their output mixes, and comparative advantages in production secured through productivity improvements and technological change. In the seventies, exchange rate adjustments account for a significant portion of the accelerated growth. The major U.S. devaluation of the early 1970s stimulated export growth. But, the trade consequences of large price changes in the world oil market are of greatest importance. Here, U.S. exporters benefitted not only from the large, new markets created in the oil-producing countries but also from the fact that world oil trade was conducted in U.S. dollars. As other oil importing nations sought to finance their petroleum imports by increasing exports and/or reducing net capital outflows, the supply of dollars increased relative to other currencies. This quickly led to movements in exchange rates. Export volumes were increased substantially as exchange rate adjustments favored the U.S. In addition, growth remained relatively strong for our industrialized trading partners, and revenues to oil- and other resource-producing nations continued to increase.

For the remainder of the century, export growth is expected to be an important component of the overall growth in final spending and output. However, the rate of growth is projected to be slower than recent experience. This results because (1) price shocks to the world oil market are assumed not to recur, (2) adjustment in exchange rates and international trade patterns in response to previous price shocks have occurred already, so that future growth in world oil prices is relatively easily accommodated, (3) future growth in the economies of our major trading partners is expected to slow somewhat compared to the 1970s, and (4) with the exception of agriculture, U.S. exporters are expected to face increased competition from foreign export producers and accelerated import substitution within importing countries.

ASSUMPTIONS FOR ADDITIONAL SCENARIOS

Besides the Reference Case, seven other scenarios were developed for analysis in this study. In the following subsections some of the major energy supply, fuel cost and economic assumptions associated with these additional scenarios are discussed.

Assumptions: Low Renewables Costs (LC) Scenario

For this scenario, the prices and costs for energy from renewables are reduced significantly, while the quantities remain at their Reference Case levels of penetration. The rate of increase in biomass prices is reduced by a full percentage point, from 3.5 to 2.5 percent per annum, resulting in a twenty percent reduction in the year 2000 from the price projected for the Reference Case.

In addition to biomass inputs, the capital and operating expenses for renewables technologies are lowered from their Reference Case values (see Table 2-5). Capital and operating costs for these technologies are reduced by an average of 10 to 40 percent. These cost reductions represent a significant cost saving to the economy for given levels of market penetration.

Reductions in renewable energy costs will have an impact on future energy use and economic output. For example, the level of demand for petroleum may be affected by any change in economic activity. To allow for such changes other variables must be permitted to adjust in the economic model. For petroleum and natural gas, only imports are permitted to adjust. By assumption, domestic production of these fuels from conventional and new sources remains at the levels specified for the Reference Case. In the electric sector, inputs of coal, oil, and natural gas to conventional technologies change in response to variations in electricity demand. The inputs to other nonrenewable electric technologies are held at their Reference Case levels. Finally, domestic coal production adjusts to changes in the quantities demanded for direct use, synthetic fuels production, and electric generation.

The final assumption, which is common to all of the additional scenarios, concerns the international aspects of changes in domestic energy demand and economic output. Changes in the value of imports (energy and non-energy alike) lead to a series of adjustments involving capital flows, exchange rates, and export volumes. U.S. experience in the 1970s indicates that large increases in oil import costs quickly led to exchange rate adjustments. These adjustments, in turn, promoted increased export and reduced import volumes and, ultimately, resulted in only minor changes in the balance of trade. The same response is assumed to occur in all of our scenarios. That is, changes in import payments are accompanied by exchange rate adjustments and by adjustments in export volumes. These adjustments are assumed to continue until the nominal (current dollar) balance of trade is restored to the same level projected in the Reference Case. This assumption does have implications for the general economic effects of energy changes. For example, to the extent that increased renewables penetration leads to a reduction in energy import expenditure, production for exports can be reduced. The resources and output thus released become available for domestic consumption and investment. Increases in energy import expenditures have the opposite effect.

Assumptions: High Nonrenewable Costs (HC) Scenario

In this scenario, the prices and costs for energy from nonrenewables are increased substantially over the levels defined for the Reference Case. Table 2-12 presents higher fuel cost assumptions. World oil price is projected to grow from $34.0 per barrel in 1980 to $86.50 per barrel by the year 2000 (1980$). This represents a 35 percent increase in the oil import price for 2000 over the Reference Case value. Following the trend in oil prices, increases in the price of natural gas are also more rapid under the higher cost conditions, reaching a level in 2000 that is 36 percent above the Reference Case natural gas price for that year. Growth in coal and nuclear fuel prices is assumed to occur at rates slightly more than one percent per year faster than estimated for the Reference Case. By 2000, coal and nuclear fuel prices are approximately 23 percent above the Reference Case levels. Increases are also assumed for the capital costs of large-scale energy production and conversion facilities (see Table 2-5).

As in the LC Scenario, assumptions regarding energy and economic adjustment mechanisms augment the energy price and technology cost changes. Petroleum imports again are assumed to adjust to demand changes, with domestic petroleum production held fixed at the Reference Case levels. Changes in demand for natural gas are accommodated by changes in both domestic production and import levels. In the electric sector, inputs to nonfossil technologies remain at their Reference Case levels so that coal, oil, and gas inputs adjust to variations in electricity demand. As before, domestic coal production changes in response to increases or decreases in demands for this fuel. Finally, export volumes adjust so as to maintain the same nominal trade balance as in the Reference Case.

Assumptions: Low Renewable and High Nonrenewable Costs (LHC) Scenario

This scenario combines the fuel price and technology cost changes of the previous two cases.

Table 2-12

ENERGY PRICE ASSUMPTIONS: HIGH NONRENEWABLES COSTS TREND
(1980 Dollars per Million Btu)

Fuel Source	1980	1985	1990	1995	2000
Crude Oil Refiner Acquisition Cost:					
Domestic	4.14	7.84	10.67	12.86	14.91
Imported	5.86	7.84	10.67	12.86	14.91
Composite	4.83	7.84	10.67	12.86	14.91
Natural Gas, Wellhead	1.49	4.12	5.73	7.05	8.35
Coal, Minemouth	1.12	1.31	1.53	1.79	2.09
Nuclear Fuel Input	0.75	0.88	1.03	1.21	1.41

Specifically, cost reductions are assumed for energy from renewable resources, while the fuel prices and technology costs for nonrenewable energy increase. The outputs from renewable technologies are held constant at the levels specified in the Reference Case. Finally, the mechanisms for economic output and energy demand adjustments in energy and the economy conform to those of the High Cost case.

Assumptions: High Renewables Penetration (HP) Scenario

This scenario is based on the expansion of renewable supplies compared to the levels introduced into the Reference Case. Higher penetration is assumed for both dispersed and centralized applications, achieving the levels shown in Table 2-13 by the end of the century. Collectively, decentralized renewables contribute an additional 1.5 quadrillion Btu (quads) of delivered energy by the year 2000 over the Reference Case (a 36.8 percent increase). Wood use in the residential sector doubles and satisfies the space-heating requirements for an equivalent of eight million households in 2000. Industrial biomass use also increases somewhat.

Although the increase in decentralized solar market penetration appears small in absolute terms, it represents a significant change in the number of units installed. By 2000, approximately 45.3 million homes, or 40.9 percent of all households, receive some fraction of their delivered energy needs from solar. (The comparable figures for the Reference Case are 31.7 million homes, or 28.6 percent of the total number of households.) The commercial sector accounts for an additional 5.8 million units of equivalent solar capacity, beyond the 13.6 million units established for the Reference Case forecast. Growth in use of solar technologies for the agricultural and industrial sectors also accelerates. Geothermal application in the industrial sector doubles as relative cost considerations are assumed to dominate the disadvantages of resource location. Finally, installed capacity of decentralized wind and photovoltaic systems is increased to 7.4 gigawatts (GWe) by 2000. This corresponds to 570,000 ten-kilowatt wind systems and 334,500 five-kilowatt photovoltaic systems, an ownership rate equivalent to 0.8 percent of all households.

Centralized use of renewables also increases. Renewable inputs in 2000 to major energy processing and conversion facilities rise from 6.8 quads in the Reference Case to 9.5 quads under higher penetration. By 2000, installed capacity from biomass and geothermal electric technologies reaches levels of 5.6 and 20.0 GWe, respectively. The increase in low-head hydro-

Table 2-13

INCREASED PENETRATION OF RENEWABLES
(Quadrillion Btu)

	2000 Level		Percent Increase[a]
	Lower Penetration Cases	Higher Penetration Cases	
Dispersed Renewables:			
Solar			
Residential/Commercial	0.700	1.000	42.9
Industrial	0.200	0.300	50.0
Biomass			
Residential	0.360	0.720	100.0
Industrial	2.800	3.500	25.0
Industrial Geothermal	0.010	0.020	100.0
Decentralized Wind	0.017	0.051	200.0
Decentralized Photovoltaics	0.007	0.010	42.9
Subtotal -- Dispersed	4.094	5.601	36.8
Centralized Renewables:			
Biomass			
Liquids	0.270	0.405	50.0
Medium Btu Gas	0.117	0.195	66.7
Electricity	0.131	0.262	100.0
Hydroelectric			
Conventional/Pumped Storage	3.703	3.703	0.0
Low-head	0.412	0.618	50.0
Geothermal Electric	1.608	3.216	100.0
Solar Thermal Electric	0.050	0.075	50.0
Centralized Wind	0.300	0.900	200.0
Centralized Photovoltaic	0.080	0.110	37.5
Subtotal -- Centralized	6.671	9.484	42.2
TOTAL	10.765	15.085	40.1

[a]Higher as compared to Lower Penetration.

electric capacity is 3.0 GWe, bringing total capacity to 9.0 GWe by the end of the century. By 2000, the rated capacities of solar thermal, wind and photovoltaic central station technologies are 2.4, 28.4, and 5.0 GWe, respectively. This implies the existence of 24 solar thermal facilities, 455 "wind farms", and 500 photovoltaic power plants. Finally, by the year 2000, biomass liquids and gases contribute the daily equivalent of 325,000 barrels of petroleum to domestic energy supplies, an increase of 54 percent over Reference Case production levels.

By 2000, overall renewables penetration is 40 percent higher under these conditions than was projected for the Reference Case. Were this increase to displace conventional energy supplies on a Btu-for-Btu basis, the share of total primary energy in 2000 provided by renewables would increase from 10.7 to 15.0 percent. However, energy changes of this magnitude affect both the economy and its use of conventional energy inputs, so this relationship is approximate, not exact.

Assumptions: High Renewables Penetration Under Alternative Cost Conditions (HPLC, HPHC, and HPLHC) Scenarios

In the remaining scenarios, the cost variations applied to the Reference Case are repeated for the higher penetration conditions. Specifically, given the increased use of energy from renewable resources, unit costs of renewable energy technologies are reduced (case HPLC), unit fuel and technology costs of nonrenewable energy are increased (case HPHC), and both renewable and nonrenewable energy costs are changed (case HPLHC). Each of these changes is defined in the preceeding subsections.

3
Economic and Energy Performance in the Reference Case

This chapter describes the economic and energy performance parameters associated with our Reference Case Scenario. The information presented in this chapter was developed using the DGEM model in conjunction with the Reference Case energy and economic assumptions described in Chapter 2.

OVERVIEW OF ECONOMIC PARAMETERS

Based on results from the DGEM model, we expect economic growth to accelerate gradually over the 1980-2000 period encompassed by our forecast. (A summary of economic parameters is presented in Table 3-1.) For the entire period real GNP growth averages about 2.9 percent per year. This growth rate is less than the 3.8 percent annual growth rate achieved from 1950 to 1973, but represents a substantial improvement over the 1973 to 1980 annual growth rate of 2.4 percent.

The rate of growth in the labor force and total employment will gradually decline over the period -- between 1980 and 1985 total employment increases at an annual rate of 1.6 percent, while for 1990 to 1995 this annual rate of increase will be only 0.9 percent. This slowed growth will occur despite increases in male and female participation rates because the working-age population will grow much more slowly during the 1980s and 1990s than during the 1960s and 1970s.

The projected increase in output per worker (a measure of gross labor productivity) over the 1980-2000 period is in the range of 1.1 to 2.0 percent per annum. This is well below the 2.2 percent annual rate of productivity advance sustained over the 1950-73 period, but considerably higher than the 0.3 percent annual rate for 1973-80. For 1980-2000, productivity growth will continue to be hampered by increases in energy prices relative to other prices. Oil and gas price decontrol and a relatively high growth rate for oil prices will impose a

Table 3-1

OVERVIEW OF ENERGY AND THE ECONOMY

	1980	1985	1990	1995	2000
Population	222.2	232.9	243.5	252.8	260.4
Real GNP	2,629.6	2,998.6	3,442.9	3,947.3	4,613.5
Real Personal Disposable Income	1,809.3	2,074.1	2,399.9	2,795.5	3,376.7
Primary Energy	77.968	83.650	88.375	94.142	100.782
Civilian Labor Force	104.8	113.2	120.2	125.4	133.0
Unemployment Rate	7.2	7.1	6.6	6.2	6.2
GNP Price Deflator	1.774	2.704	3.846	5.303	7.175
Energy-GNP Ratio	29.7	27.9	25.6	23.9	21.8

AVERAGE ANNUAL GROWTH RATES

	1980-85	1985-90	1990-95	1995-2000
Population	0.92	0.89	0.75	0.59
Real GNP	2.66	2.80	2.77	3.17
Real Personal Disposal Income	2.77	2.96	3.10	3.85
Primary Energy	1.42	1.10	1.27	1.37
Employment	1.57	1.32	0.93	1.20
Gross Labor Productivity	1.07	1.47	1.83	1.95
GNP Price	8.80	7.30	6.64	6.23
Energy-GNP Ratio	-1.21	-1.67	-1.45	-1.71

Units:

Population and Civilian Labor Force in millions of persons.
Real GNP and Real Personal Disposal Income in billions of 1980 dollars.
Primary Energy in quadrillion Btu.
Unemployment Rate in percent.
GNP Price Deflator based on 1972 = 1.0.
Energy-GNP Ratio in thousands of Btus per 1980 dollar.
Growth Rates in percent.

substantial productivity cost on the economy during the early and middle 1980s. However by the late 1980s and during the 1990s substantial improvements in productivity growth should occur. This improvement will be assisted by a labor force characterized by higher skill levels as the average age of workers increases.

Projected increases in gross labor productivity include the effects of increased labor efficiency, more capital per worker, improved capital efficiency, and changes in the sectoral mix of production. Increases in capital per worker are particularly important in advancing labor productivity. The ratio of capital input to labor input is estimated to increase more than 50 percent between 1980 and 2000. This trend reflects, among other things, the relative price implications of both the changing population age structure (which is likely to increase wage rates relative to other prices) and the favored tax treatment afforded to future capital accumulation (which is likely to lower capital prices relative to other prices).

Growth in energy consumption exhibits marked changes from previous trends. Future growth in energy consumption is likely to be more rapid than the 0.2 percent annual rate since 1973 because the large and sudden price shocks of the 1970s should not be repeated. Projected patterns of energy growth reflect lagged responses to price and regulatory changes occurring in the 1970s and over the 1980 to 2000 forecast period. Overall, the economy's use of energy will become progressively more efficient over time. Thus, the energy-GNP ratio, which declined at an average annual rate of 0.1 percent annually (despite falling relative energy prices) between 1950 and 1973 and at a 2.1 percent annual rate between 1973 and 1980 (when substantial increases in real energy prices occurred), is projected to decline at an annual average 1.5 percent during 1980 to 2000 (a period that will be characterized by gradual increases in energy prices).

Reducing the energy intensity of the economy or, equivalently, increasing the gross economic efficiency of energy use, depends on increases in the technical efficiency of energy conversion and end use, technical progress in the economy as a whole, and structural changes in spending, input, and output patterns. Changes in the pattern of energy use are conditional on the relative effects of these influences. For the 1980s, the dominant cause of improved energy efficiency is the economy's adjustment to the rapid increase in world oil prices (1979-1985), deregulation of domestic petroleum prices (1981), and partial decontrol of natural gas prices (1985). Energy policy measures initiated in the 1970s (such as tax credits for energy conservation expenditures) also contribute to the increased efficiency of energy use. For the 1990s, structural changes in spending, input, and output patterns (caused in part by rising energy prices), continuing increases in energy costs, and improving productivity account for reductions in the energy intensity of the economy. On balance, these energy changes indicate the economy's ability to accommodate more costly and less readily available energy supplies without corresponding reductions in the growth of real output and incomes.

Disposition of Final Output

The increasing quantity of production permits a continuing increase in material standards of living. Real disposable income per capita increases by 59 percent from 1980 to 2000, from $8,100 to $13,000. Gross national product (GNP) increases by 74.4 percent over this same period. In Table 3-2, GNP is separated into purchases by consumers, businesses, governments, and others. Personal consumption expenditures (PCE) remain the dominant use of production (60 to 65 percent of total output) as a result of two important factors. First, consumption absorbs much of the impact of rising energy prices. PCE growth is slower than real GNP growth during the 1980s, reflecting the effects of relatively high oil price growth and natural gas price decontrol, and is faster than real GNP growth over the 1990s, when energy price growth is moderated. Second, and perhaps more important, changes in the tax structure serve to stimulate saving and investment at the expense of current consumption.

On the use side, investment as a share of real GNP gradually increases. By contrast, the fraction of total final output going to government falls from current levels of over nineteen percent to 16.6 percent by 2000. Reductions in the growth of government purchases permit a larger portion of private saving to be absorbed by business and household investment. Finally, foreign trade is projected to continue to increase in importance over the remainder of the century. Exports become an increasingly important use of final output, even though export growth occurs at rates well below recent historical levels. Total national expenditure for petroleum imports remains quite high through the 1980s and 1990s.

Patterns of Relative Price Changes

Prices of various goods and services increase at rates reflecting movements in production costs, input

Table 3-2

DISPOSITION OF TOTAL FINAL OUTPUT

	1980	1985	1990	1995	2000
Purchases, Billions of 1980$					
Consumption, C	1,657.7	1,843.5	2,142.9	2,481.4	3,024.1
Investment, I	362.9	479.8	568.1	690.8	784.3
Government, G	514.8	579.5	647.3	706.4	767.1
Exports, X	286.7	337.1	393.2	462.1	557.2
Imports, M[a]	192.5	241.3	308.6	393.4	519.2
Petroleum Imports	79.1	107.6	115.9	117.1	97.6
GNP	2,629.6	2,998.6	3,442.9	3,947.3	4,613.5
Composition of Purchases, Percent					
Consumption, C	63.0	61.5	62.2	62.9	65.6
Investment, I	13.8	16.0	16.5	17.5	17.0
Government, G	19.6	19.3	18.8	17.9	16.6
Exports, X	10.9	11.2	11.4	11.7	12.1
Imports, M	7.3	8.0	8.9	10.9	11.3
Petroleum Imports	3.0	3.6	3.4	3.0	2.1
GNP	100.0	100.0	100.0	100.0	100.0

Note: GNP = C + I + G + (X-M)
Percentages may not add to 100 % due to rounding.

[a]Includes petroleum imports. Petroleum imports are also noted separately for informational purposes.

prices, and sectoral productivity. Table 3-3 presents the annual rates of change in the prices of each of the major categories of finished goods and services and for the primary inputs to production.

Energy prices are projected to increase more rapidly than all other input prices. These price changes reflect the rise in world oil prices, oil and gas price decontrol, and continuing cost increases in energy processing and conversion. Labor prices also are projected to increase relatively quickly. For workers, a rising relative price of labor implies an increase in purchasing power and material standards of living. For producers, increasing relative labor prices raise production costs and provide an incentive for substitution away from the labor input. Since capital and materials prices increase at less rapid rates, the control of production costs, i.e., cost minimizing behavior, implies a restructuring of input patterns toward these factors.

A change in the relative price structure also is projected for the output side of the economy. Since energy input costs are rising most rapidly, prices of energy output show the most rapid annual rate of increase. Overall, there is a wide range of movement in relative prices. For example, energy prices increase much faster (about 3.8 percent per year) than services prices over 1980-2000. These differentials exert substantial influence on the patterns of final spending, intermediate purchases, and sectoral production, and, thus, the structure of economic growth.

Structure of Final Spending

The composition of final demand spending is projected to change materially over time in response to changes in relative prices. Final demand spending covers personal consumption expenditure, investment, government purchases, and imports. The allocation of final demand among principal categories of goods and services is shown in Table 3-4. The structure of this spending changes over time in response to several influences.

The most striking change is the reduced role of energy in the volume of final spending, which falls continuously from 4.4 percent of total purchases in 1980 to 2.8 percent by the year 2000. Much of the reduction in energy purchases occurs because households adjust their expenditure patterns in response to increases in relative energy prices.

Agriculture, nonfuel mining, and construction also decline in relative importance from 1980 to 2000. Reductions in the demands by households (as incomes and

Table 3-3

OUTPUT AND INPUT PRICE INCREASES
(Average Annual Rates of Change in Percent)

	1980-85	1985-90	1990-95	1995-2000
Final Output Prices				
Agriculture, Nonfuel Mining, Construction	9.82	7.74	6.99	6.29
Nonenergy Manufacturing	8.83	7.40	6.89	6.48
Transportation	7.97	7.04	6.70	6.55
Communications, Trade, Services	7.65	6.75	5.91	6.03
Energy	13.31	10.60	9.68	9.02
Average	8.80	7.30	6.64	6.23
Input Prices				
Capital	8.16	6.23	5.78	5.06
Labor	9.96	8.88	8.59	8.30

Table 3-4

COMPOSITION OF REAL FINAL DEMAND
(Percent of Total Constant Dollar Final Demand)

	1980	1985	1990	1995	2000
Agriculture, Nonfuel Mining, Construction	13.8	13.6	13.4	13.2	12.9
Nonenergy Manufacturing	35.7	35.9	35.7	35.5	34.9
Transportation	3.5	3.5	3.5	3.6	3.7
Communications, Trade, Services	42.6	43.0	43.8	44.5	45.7
Energy	4.4	4.0	3.6	3.2	2.8
TOTAL	100.0	100.0	100.0	100.0	100.0

relative prices change) and by governments (due to budgetary restrictions) for goods from these sectors are the major factors for the decreasing importance of these sectors.

The share of nonenergy manufacturing in total final spending is stable through the 1980s and declines only slightly during the final decade of the century. There are important causes for the stability of manufacturing's share. In the household sector, increases in relative energy prices are larger than those for manufactured goods. Consequently, there is limited substitution of these goods for energy. In addition, manufacturing's share of total business purchases is relatively large so that the shift toward investment tends to stabilize its overall relative importance. The restructuring of government spending toward defense reinforces this effect. Finally, manufacturing continues to play an important role in international trade, despite increasing competition from foreign producers.

The remaining sectors (commercial transportation and communications, trade, and services) absorb an increasing fraction of the quantity of total spending. Here, relative price considerations and the systematic shift in expenditure and preference patterns are the principal causes.

Structure of Inputs to Production

The pattern of inputs used in domestic production also changes considerably over time. Table 3-5 shows the constant dollar proportions of total input for capital, labor, energy, and intermediate materials. There are significant changes in the relative importance of each of these aggregate inputs. Production becomes progressively more capital intensive and less labor intensive, continuing the long-run historical trend. The energy intensity of production increases from 1980 to 1985 but declines continuously thereafter, and the relative importance of materials inputs rises steadily throughout the projection period. With the exception of energy, these changes conform exactly to expectations regarding the effects of relative price changes for inputs (Table 3-3).

The energy intensity of production merits special discussion. As given by the constant dollar proportion of total input, energy rises from 5.7 to 6.2 percent between 1980 and 1985 and, then, declines only gradually to 5.9 percent of total inputs by the end of the century. This appears to be inconsistent with both the pattern of rising relative input prices for energy and the declining levels of energy use, measured in physical units per dollar of output, within the economy (see

Table 3-5

AGGREGATE INPUT PATTERNS
(Share of Total Constant Dollar Inputs to Production)

	1980	1985	1990	1995	2000
Input Share for					
Capital	0.1400	0.1430	0.1488	0.1528	0.1595
Labor	0.2762	0.2611	0.2440	0.2240	0.2061
Energy	0.0574	0.0617	0.0614	0.0606	0.0593
Materials	0.5264	0.5342	0.5458	0.5626	0.5751

Annual Percentage Growth Rates	1980-85	1985-90	1990-95	1995-2000
Capital per Unit of Labor	1.56	2.17	2.27	2.56
Energy per Unit of Capital	1.03	-0.89	-0.79	-1.28

Tables 3-1 and 3-3). However, there are several important mechanisms underlying these energy changes. Energy inputs are measured in constant dollars per physical unit, i.e., Btu, of purchased energy. In constant dollars per unit, electricity is most expensive, followed, in descending order, by refined petroleum products, delivered gas, and coal. In terms of relative price increases within energy, gas prices rise most rapidly followed by the escalations in petroleum, coal, and electricity prices. With this pattern of relative energy price changes, there occurs a shift in energy inputs away from oil and gas and toward electricity and, to a lesser degree, coal. Also, in response to changes in spending and interindustry expenditure patterns (Tables 3-4 and 3-6, respectively), there occurs a shift in U.S. production toward the communications, trade, and services industries (Table 3-7). Here, electricity for gas substitution dominates other adjustment mechanisms, amplifying the overall shift toward electricity.

Since electricity is the least efficient form of energy (in terms of the ratio of Btu output to Btu input), the time pattern of energy shares disguises a continuing and significant decline in delivered energy use per unit of output. Since electricity is also most expensive "fuel" in real terms, the dollar share of energy per unit output increases initially.[1] In the short-run (1980-1985), reductions in the growth of energy use in production are not sufficient to prevent the rising energy share that results from the shift toward electricity. After 1985, reductions in energy demand growth more than compensate for the shift toward electricity, resulting in a decline in energy's share of inputs to production.

<u>Structure of Domestic Demand and Supply</u>

The previous two sub-sections discussed the structure of economic growth from the perspective of intermediate and final purchasers. This subsection discusses the aggregate of intermediate and final demand (or total demand) and total supply.

Table 3-7 shows the structure of total domestic production for the 1980 to 2000 period. Several

---

[1]Note that energy's share of total final spending decreases continuously implying that the rate of electricity demand growth in the household sector declines significantly in the 1980s and 1990s as compared to the 1960s and early 1970s.

Table 3-6

COMPOSITION OF REAL INTERMEDIATE PURCHASES
(Percent of Total Constant Dollar Interindustry Expenditure)

	1980	1985	1990	1995	2000
Agriculture, Nonfuel Mining, Construction	7.1	6.8	6.8	6.7	6.9
Nonenergy Manufacturing	49.4	48.4	48.2	47.9	48.2
Transportation	5.5	5.5	5.6	5.7	5.8
Communications, Trade, Services	28.2	28.9	29.3	30.0	29.8
Energy	9.8	10.4	10.1	9.7	9.3
TOTAL	100.0	100.0	100.0	100.0	100.0

Table 3-7

COMPOSITION OF REAL DOMESTIC OUTPUT
(Percent of Total Constant Dollar Production in the United States)

	1980	1985	1990	1995	2000
Agriculture, Nonfuel Mining, Construction	10.6	10.5	10.4	10.4	10.4
Nonenergy Manufacturing	42.1	41.5	41.0	40.5	40.0
Transportation	4.3	4.2	4.2	4.2	4.1
Communications, Trade, Services	36.3	37.0	37.8	38.6	39.5
Energy	6.7	6.8	6.6	6.3	6.0
TOTAL	100.0	100.0	100.0	100.0	100.0

differences emerge in comparing this pattern of outputs (domestic supply), with those for final demand and intermediate purchases (total demand). The supply and demand trends in agriculture, nonfuel mining, and construction are virtually identical, implying that the economy remains almost entirely self-sufficient in these output categories. Also, over time, the outputs from these sectors decrease as a proportion of total economic output.

Total outputs in the communications, trade, and services industries become progressively more important over time as the economy becomes increasingly more services-oriented in demand and supply. Again, imports contribute only a small share of total supply as growth in domestic production keeps pace with demand growth.

In the remaining sectors, however, domestic production grows at different rates from total demand. Energy's share of total domestic output is stable through the mid-1980s and declines only gradually thereafter. From 1980 to 1985, slower growth in energy demand is accompanied by a shift away from oil and gas and toward electricity and coal. However, these changes are not sufficient to prevent an increase in oil and gas imports for this period. After 1985, there are significant reductions in the share of energy within total demand while, in production, the decrease in energy's relative importance is much less dramatic. This pattern reflects the increasing contribution of domestically produced energy (coal, electricity, synthetic fuels, renewables, etc.) and the diminishing role of imported energy in overall supply. Thus, there is not only a shift away from energy in response to price changes and changes in input and spending patterns, but also a gradual shift away from imports in energy supply.

Demand trends in nonenergy manufacturing are marked by stability, both in final spending and as a productive input. Yet, there is a continuous decline in its relative importance within total domestic output. This implies that imports of manufactured goods increase in relative importance. In fact, the import share of total supply for nonenergy manufacturing rises from 3.0 percent in 1980 to 7.6 percent by the year 2000.

Commercial transportation increases in relative importance within final spending and intermediate purchases. However, its share of total U.S. production is stable throughout the projection horizon. Like manufacturing, this implies a significant increase in the use of imported transportation services.

The production trends in manufacturing and commercial transportation have important implications for future energy demand. Growth in energy use is slowed both in direct response to energy price increases and, indirectly, through effects on the structure of final spending and intermediate purchases -- i.e., less energy intensive goods and services are demanded. Also, both manufacturing and commercial transportation are relatively energy intensive, so that growth in energy demand is slowed further by importing it in the embodied form of finished goods and transportation services. This suggests that some portion of the energy import problem confronting the U.S. is exported to other nations.

Summary of Reference Case Economic Growth

The main characteristics of the U.S. economy over the remainder of the century for the Reference Case can be summarized as follows:

- Real GNP growth is relatively low during the early 1980s reflecting slower employment growth (compared to the 1970s), a continued low rate of productivity advance, and higher energy prices;
- Real GNP growth is higher over the period 1985 to 1995. The moderation of energy price increases and increasing productivity growth compensate for the continuing slower growth of the labor force;
- From 1995 onward, real GNP growth is sharply increased as labor force expansion accelerates and augments continuing productivity improvements;
- Productivity growth accelerates over time;
- Consumption remains the dominant use of final output, with investment, exports, and imports increasing and government purchases declining in relative importance. Tax incentives that favor saving and investment and the reduction in the growth of government expenditure all help stimulate capital formation.
- Relative prices change significantly. For inputs, the prices of energy and labor rise relatively more rapidly than do those for capital and materials. For outputs, the prices of energy and energy-intensive goods and services rise more rapidly than those of the services-oriented sectors;
- Within final spending, there is a structural shift away from energy and energy-intensive goods and services. Final spending is redirected toward the outputs of the services industries. These changes occur in response to relative price changes in the disposition of final output.

- Production becomes increasingly more capital intensive and less labor and energy intensive. There is, however, a shift toward relatively greater use of electricity. Also, the relative importance of materials, as an input to production, increases. This implies that production becomes more intensive in purchased goods and services. Finally, the economy becomes progressively more service-oriented.
- Growth in domestic supply conforms to demand growth in all sectors except energy, manufacturing, and commercial transportation. In energy, domestic production gradually displaces imports. For manufacturing and transportation, imports account for an increasing fraction of total supply. Growth in energy demand is slowed not only by structural changes in final spending and intermediate purchases but also by shifting the production of energy-intensive goods and services to foreign producers.
- The shifts in spending, input and output patterns permits a continuous decline in the energy intensity of the economy.

PATTERNS OF ENERGY USE AND SUPPLY IN THE REFERENCE CASE

Delivered Energy Use

Delivered energy is energy consumed by nonenergy producers and final demand sectors. It excludes all energy used or lost in primary energy conversion,[2] processing, and distribution activities. Use of energy in production and final demand is projected to increase over time. However, substantial changes occur in fuel-use patterns and, in response to relative price changes, the overall growth in delivered energy use is well below historical trends. The emerging patterns of energy use reflect compositional changes in output and demand, input adjustments, and interfuel substitutions. Table 3-8 presents information on and growth of delivered energy consumption for the 1980 to 2000 period.

Total use of delivered energy increases from 56.5 quadrillion Btu (quads) in 1980 to 69.2 quads by the year 2000, an average annual growth rate of 1.0 percent. Since real GNP rises at a 2.9 percent annual rate

---

[2]Thus, energy lost through conversion or transportation activities prior to the delivery of fuel to final demand or an intermediate producer are excluded from delivered energy.

Table 3-8

DELIVERED ENERGY CONSUMPTION
(in Btus and Percentage Terms)

	1980	1985	1990	2000
Consumption, quadrillion Btu				
Coal	3.489	4.434	5.543	8.580
Petroleum	28.450	29.380	27.920	26.710
Gas	15.575	15.460	16.233	16.974
Electricity	7.136	8.475	9.873	12.793
Dispersed Renewables[a]	1.809	1.986	2.257	4.094
TOTAL	56.459	59.735	61.826	69.151
Consumption in Percent				
Coal	6.18	7.42	8.97	12.41
Petroleum	50.39	49.18	45.16	38.63
Gas	27.59	25.88	26.26	24.55
Electricity	12.64	14.19	15.97	18.50
Dispersed Renewables[a]	3.20	3.33	3.64	5.91
TOTAL	100.00	100.00	100.00	100.00

[a] Includes solar thermal (residential, commercial, industrial), biomass (residential, industrial), industrial geothermal, decentralized wind, and decentralized photovoltaics. See Table 2-10 for details.

over this period, delivered energy use per constant dollar of output declines at an average annual rate of 1.8 percent, yielding a 30.1 percent reduction in average delivered energy requirements per dollar of real GNP. This reduction is concentrated in the 1980 to 1990 period, reflecting a lagged response to relatively rapid oil price growth (1979-1985), oil and gas price decontrol (1981 and 1985, respectively), and continuing increases in the cost of energy.

Shifts in the mix of fuels used are also very important. Use of petroleum peaks in the mid-1980s, gradually declining thereafter. This reflects the combined influences of price increases and regulatory changes. Relatively rapid increases in petroleum prices during the early 1980s, together with oil price decontrol, will encourage conservation of petroleum in all end uses. Regulations, such as fuel-efficiency standards for automobiles, reduced speed limits, fuel-backout provisions, will add to the incentives to economize on petroleum.

Natural gas use falls slightly between 1980 and 1985. In part, growth in gas consumption is supply constrained as a result of the continuation of current regulatory policies in gas supply and use. However, the primary reason for this demand reduction is price. Natural gas prices rise rapidly from 1980 to 1984, under the provisions of the Natural Gas Policy Act, and increase sharply in 1985, as full decontrol occurs for many categories of gas supply. As with petroleum, these price changes promote conservation in all phases of gas use. After 1985, natural gas price increases should moderate significantly and use increase slightly.

Sustained growth in the use of coal, electricity, and dispersed renewables will occur over the entire forecast period, and these fuels will substitute, to some extent, for petroleum and gas. Direct use of coal is projected to more than double despite environmental and capital cost disadvantages, as coal fuel becomes increasingly inexpensive relative to other fuels. Electricity consumption also increases at a relatively rapid rate, averaging 3.0 percent per year. This is well below historical rates of growth, reflecting the general trend toward lower growth in energy demand. Even though electricity is more expensive on a delivered Btu basis than other fuels, its ease and flexibility of use, and relatively low rate of price increase (compared to other fuels) allow it to increase its share of total delivered energy. Also, the fastest growing sectors of the economy, such as commercial and service activities, are relatively intensive users of electricity. Finally, use of dispersed renewables to satisfy delivered energy needs is assumed (exogenously) to rise significantly through the year 2000. (Total energy demand is projected endogenously using DGEM.)

Electricity Generation

Electricity is projected to play an increasingly important role in the nation's overall energy system. Consumption of electricity should increase at an average annual rate of 3.0 percent over the 1980 to 2000 period, compared to 2.9 percent growth in the economy and only 1.0 percent growth in total delivered energy demand. The share of electricity in delivered energy rises from 12.6 percent to 18.5 percent while the fraction of primary energy input used for electricity generation shows a corresponding increase from 32.1 percent to 41.0 percent. Of the three current principal forms of delivered energy (i.e., oil, gas and electricity), electricity is the only one for which significant demand growth occurs.

The mix of energy sources used for electric power generation is summarized in Table 3-9. In 1980, coal provides almost half of the total input with petroleum, natural gas, nuclear, and renewables inputs each providing from 10.0 to 15.0 percent of the total. This mix will change significantly over the remainder of the century. Use of petroleum and natural gas will be reduced significantly. This decline is due to a shift by utilities away from oil and gas in response to both the rapid price increases for these fuels and government regulatory policies designed to accelerate the substitution process. Coal, nuclear, and renewables inputs all increase in absolute terms but the largest increase occurs in coal use. Coal input grows at 3.6 percent annually and provides 59.3 percent of total electric inputs by the turn of the century. Nuclear's share increases from 10.9 percent in 1980 to 20.2 percent by 2000 while the share of total inputs provided by renewables rises from 12.9 to 15.2 percent.

Primary Energy Consumption[3]

Aggregate U.S. primary energy consumption for the Reference Case is projected to increase from 78.0 quads in 1980 to 100.8 quads by the year 2000. This represents a 29.3 percent increase over present levels

---

[3]Primary energy consumption includes all energy consumed, including energy lost through conversion (e.g., to electricity of synthetic fuel) and transportation (i.e., electric transmission-related losses).

Table 3-9

ELECTRIC GENERATION
(Quadrillion Btu)

Central Station Inputs	1980	1985	1990	1995	2000
Coal:					
Conventional	10.284	12.318	12.539	11.336	6.783
Advanced[a]	1.815	3.521	6.159	10.809	17.725
Biomass	0.010	0.020	0.058	0.087	0.131
Oil	3.170	2.300	1.650	1.100	0.900
Gas	3.812	2.822	2.186	1.693	1.312
Nuclear	2.727	4.560	6.560	7.440	8.330
Hydroelectric:					
Conventional/ Pumped Storage	3.101	3.160	3.349	3.509	3.703
Low-Head	0.001	0.040	0.100	0.200	0.412
Geothermal	0.108	0.389	0.624	1.002	1.608
Solar Thermal	--	--	0.003	0.006	0.050
Wind	--	0.001	0.010	0.100	0.300
Photovoltaics	--	--	0.005	0.015	0.080
TOTAL	25.028	29.131	33.243	37.297	41.334
TOTAL GENERATION	7.874	9.351	10.894	12.476	14.116
DELIVERED ELECTRICITY	7.136	8.475	9.873	11.307	12.793

[a]Includes coal facilities equipped with scrubbers and cooling towers, coal combined-cycle systems, fluidized bed combustion, and coal-based cogeneration units.

or annual growth rate of 1.3 percent. Growth in primary energy use is moderate through both the 1980s and 1990s. The higher relative growth in primary as compared to delivered energy results from the increased use of energy provided by indirect means, i.e., electricity generation and synthetic fuels production. Within the overall demand for energy, there will be significant changes over time in the mix of resources used (see Table 3-10). Direct coal use and coal inputs to the electric and synthetic fuels industries account for a major portion of the overall increase. Total use of renewable resources also more than doubles, increasing from 5.0 quads in 1980 to 10.8 quads by 2000. Increased utilization of renewables technologies occurs in decentralized applications, electric generation, and synthetic fuels production (alcohol fuels and gas from biomass). Growth in utilization of nuclear energy also is projected to be relatively rapid, averaging 5.7 percent annually over the 1980 to 2000 period. This growth, though large, merely reflects the scheduled completion of the nuclear power plants currently under construction or in the advanced planning stages.

In contrast to these, use of conventional oil and natural gas resources steadily declines over time. These reductions reflect the effects of prices and regulatory efforts in curtailing oil and gas consumption. However, output from synthetic fuels production increasingly augments conventional liquids and gas supplies, so that these primary input reductions overstate the substitution away from delivered petroleum and gas use. Nevertheless, the oil and gas reductions are significant -- conventional oil and gas provide only 44.0 percent of total primary input in 2000 compared to 70.1 percent in 1980.

Oil Imports and Trade

Domestic oil production is projected to fall from present levels of 10.2 million barrels a day (MMBD) to 8.0 MMBD in 1995 and, then, to increase to 8.5 MMBD by the end of the century. The reversal in production trends after 1995 reflects the assumption of successful development of frontier areas in Alaska and the lower-48 and large-scale use of enhanced-recovery techniques.

The difference between demand and domestic supply must be met by imported petroleum. Thus, oil imports increase slightly from 6.4 MMBD (million barrels per day) in 1980 to 7.4 MMBD in 1985, declining steadily from 1985 onward to about 4.2 MMBD in 2000. Consequently, "oil import vulnerability" continues to present a potential problem for the U.S. well into the 1990s.

Table 3-10

PRIMARY ENERGY USAGE
(Quadrillion Btu)

	1980	1985	1990	1995	2000
<u>Nonrenewables</u>					
Oil: Domestic	20.500	17.900	15.900	15.900	16.900
Imported	13.500	15.570	14.198	12.240	8.893
Shale	--	0.090	0.290	0.460	0.730
Gas: Domestic	19.655	17.650	18.520	17.340	17.130
Imported	0.969	1.761	0.840	1.477	1.421
Coal	15.588	20.465	25.528	31.189	36.613
Nuclear	2.727	4.560	6.560	7.440	8.330
Subtotal	72.939	77.996	81.836	86.046	90.017
<u>Renewables</u>					
Biomass	1.810	2.018	2.271	2.834	3.678
Hydroelectric	3.102	3.200	3.449	3.709	4.115
Geothermal	0.108	0.389	0.628	1.009	1.618
Solar	0.009	0.044	0.166	0.413	0.950
Wind	--	0.003	0.019	0.113	0.317
Photovoltaics	--	--	0.006	0.018	0.087
Subtotal	5.032	5.654	6.539	8.096	10.765
TOTAL	77.968	83.650	88.375	94.142	100.782
Percent Renewables	6.5	6.8	7.4	8.6	10.7
Percent Imports	18.6	20.7	17.0	14.6	10.2

A relatively steady level of oil imports in conjunction with rising world oil prices means that the nominal bill for U.S. oil imports increases over time. Oil import payments are projected to increase from $79 billion in 1980 to $395 billion by 2000 (in current dollars). However, in the long run, oil imports will diminish as a percentage of total import payments (nominal and real).

Summary of Reference Case Energy Growth Projections

The main features of the U.S. energy system for the Reference Case scenario over the 1980 to 2000 period can be summarized as follows:

- Delivered energy use and primary energy inputs continue to grow, but at rates of growth well below historical levels and below the projected rates of growth for production and spending in the economy;
- Use of petroleum and natural gas steadily decrease in relative importance within the energy system;
- Coal, electricity, and renewables account for the growth in delivered energy;
- Coal, nuclear, and renewables account for additions to electric generation capacity;
- Petroleum import requirements diminish in the long run, after a small increase in 1980-85.

4
Energy and the Economy: Effects of Changes in Energy Conditions

Just as the economic and energy forecasts for the Reference Case were described in Chapter 3, this Chapter presents economic and energy forecasts for each of the other seven scenarios developed for this study. The results in Chapters 3 and 4 are then compared in the next Chapter.

LOW RENEWABLE ENERGY COSTS (LC) CASE

In this scenario year 2000 renewable energy technology costs were reduced from ten to forty percent (depending on the particular technology) below their Reference Case levels (see also Table 2-5). Since the Reference Case costs for these technologies represent the most likely costs, this Case implicitly assumes that additional cost reductions are secured through additional technological or production method advances beyond those assumed for the Reference Case. Cost savings (compared to the Reference Case) for biomass fuels were also assumed for this case through a reduction in their real rate of increase.

Reducing the costs of renewable technologies from those assumed for the Reference Case has a favorable, but modest, impact on the growth and performance of the U.S. economy. These effects are summarized in Table 4-1. Lower renewable costs lead to more rapid growth in overall economic productivity and productive capacity. Thus, higher levels of output and real income can be achieved. Also, more rapid productivity growth permits a slight reduction in the rate of overall price inflation. However, overall energy use is increased slightly, in line with higher incomes and spending.

None of these changes is large. Compared to the Reference Case projection, real annual Gross National Product (GNP) growth is increased by a mere 0.02 percentage points. By 2000, real GNP is only $13.1 billion

Table 4-1

OVERVIEW OF ENERGY AND THE ECONOMY:
THE LOW RENEWABLE ENERGY COSTS (LC) CASE

	1980	1985	1990	1995	2000
Real GNP	2,630	3,000	3,446	3,954	4,627
Primary Energy Use	78.0	83.7	88.4	94.2	100.9
Energy-GNP Ratio	29.7	27.9	25.7	23.8	21.8

AVERAGE ANNUAL GROWTH RATES, 1980-2000

	Reference Case	LC Case
Real GNP	2.85	2.87
Primary Energy Use	1.29	1.30
Energy-GNP Ratio	-1.51	-1.52
Productivity	1.58	1.59
Inflation	7.24	7.23
Real Consumption	3.05	3.06
Real Investment	3.93	3.98
Capital-Labor Ratio	2.14	2.16

Units:
Real GNP in billions of 1980$.
Primary Energy in quadrillion Btu.
Energy-GNP Ratio in thousands of Btus per 1980$.
Growth Rates in percent per annum.
Productivity is real GNP per person employed.
Inflation is the rate of growth in the GNP price deflator.

(1980$), or 0.28 percent, higher under the lower renewable costs. Annual productivity growth increases and inflation decreases by 0.01 percentage points, respectively. Finally, the increased usage of primary energy is proportionally less than the increase in economic output, so there occurs a small increase in the overall economic efficiency of energy use.

The energy-economy interactions that underlie these changes are complex, but several important mechanisms can be identified. The lower costs for renewable energy supplies mean that the same amount of delivered energy services now can be produced for a smaller commitment of capital, labor and other economic resources. These lower costs permit a release of capital, labor, and other inputs for use in other applications. The use of these economic resources in other areas permits total production to increase. Incomes and purchasing power also increase. Further, since the greater output is achieved for a given volume of capital and labor input, the average productivity of the economy rises.

The volume of expenditures grows in accordance with the sustained increases in the quantity of production. All purchases except those by the government sector share in this increased growth (by assumption, the volume of government purchases was kept the same in all scenarios as in the Reference Case).[1] Real personal consumption, out of higher real incomes, is slightly higher than in the Reference Case. In relative terms, investment shows the largest increase. There are several reasons for this rise in investment. Higher incomes and rates of return lead to some rise in private saving, thereby increasing the supply of resources for investment. Since economic output is higher, government revenues are increased over the Reference Case levels. Also, public outlays fall slightly as a proportion of GNP compared to the Reference Case, because incomes are slightly higher and prices are slightly lower. Further, the reduced unit cost of renewable energy means that fewer private and public financial resources are

---

[1]Thus, government expenditures for goods and services do not vary across the scenarios but certain transfer payments may vary. The most important of these variable transfers are the subsidies offered by the public sector (mainly the federal government) to renewable energy users. These subsidies are assumed to occur as long as the total costs of obtaining energy for any renewable technology are higher than the cost of energy obtainable from a comparable nonrenewable technology.

required to support these energy activities. Less government finance occurs in the form of either lower outright payments (grants) or smaller tax revenue reduction (tax incentives); less private finance occurs as costly renewable technologies absorb fewer of the economy's financial resources. Together, these changes result in a reduced government deficit, allowing more savings to flow into private investment. In turn, higher investment leads to more rapid growth of productive capacity which, in part, underlies the higher levels of production and real incomes.

Cost reductions for renewable supplies have a small impact on the structure of relative prices, although the trends from the Reference Case are preserved. The release of resources from the production of renewable energy supplies and their simultaneous reallocation to other uses, improves both sectoral and overall productivity growth. This has a favorable impact on production costs and permits slightly slower growth of intermediate and final output prices. Also, lower costs for renewables reduce the annual rate of growth in the average price of energy, even though use of high-cost conventional supplies increases. These adjustments lead to a slight reduction in the rate of growth of prices for agricultural, construction, transportation and sevices products while prices for manufactured goods are largely unaffected. Overall, there is a small reduction in the rate of inflation.

The increase in real incomes and the small change in the price structure leads to some change in the pattern of final demand spending (i.e., spending for consumption, investment, government and export purposes on the various types of goods and services). In addition, the rise in investment spending and the fall in government purchases, relative to the other categories of spending, leads to a change in the composition of final demand purchases. The share of spending devoted to construction and manufacturing rises while energy purchases retain their share, and the volume of transportation and services purchases declines relative to total demand.

The pattern of inputs to production shows more marked adjustments. Through 1985, the input structure is identical to that of the Reference Case. But, after 1985 as the reductions in renewables costs become more significant, there are more pronounced changes in input patterns. The energy-intensity of production is unchanged while the share of intermediate materials used per dollar of output is increased. But the more important changes involve capital. The rise in investment permits a relative rise in capital intensity of production. Thus, there is a more rapid growth in capital per unit of labor than in the Reference Case and a further decline in the ratio of energy to capital input. Otherwise the structure of intermediate output and puchases remains virtually unchanged. Thus, it is the change in capital availablility and use that has the principal impact on input structure, output potential and productivity change.

As economic growth is increased under conditions of lower renewables costs, so too is the economy's use of energy. Since the quantity of energy from renewables sources is fixed (by assumption) in this scenario, the additional energy requirements are provided from conventional sources. Table 4-2 compares energy consumption patterns for the year 2000 for the Reference Case and LC scenarios. Total use of delivered energy in 2000 is increased by slightly more than 0.1 quadrillion Btu (quads), or 0.15 percent. Since this increase is proportionately less than the increase in real economic activity, the economic efficiency of energy use rises by 0.2 percent. All conventional inputs share in the increase in overall energy demand. The largest quantity increases occur in use of oil and gas, with some rise in electricity consumption. The increase in electricity demand is supplied mainly by increased use of coal and, to a lesser extent, by additional inputs of oil and gas above their Refernce Case levels. Finally, with domestic oil and gas production held at their Reference Case levels, oil and gas imports rise to satisfy the additional demand for these fuels. In terms of primary energy use, coal accounts for the largest share of the overall increase (43 percent) followed by imported gas (30 percent) and imported petroleum (27 percent).

In summary, cost reductions for renewable energy technologies (and biomass fuels) enable economic resources to be released for other, more productive uses. This results in an increase in economic output, which, in turn, causes a small, but noticeable increase in overall energy use. Slightly more than one-half of this increase in energy use is supplied by import of oil and natural gas. Thus, an indirect effect of reducing the costs of energy from renewable sources is an increase, albeit small, in the use and importation of conventional fuels.

HIGH NONRENEWABLE ENERGY COSTS (HC) CASE

In this case, costs for nonrenewable fuels (oil, coal, natural gas and uranium) are increased significantly over their Reference Case levels. Natural gas and petroleum prices for the year 2000 are more than one-third higher than for the Reference Case, while uranium and coal costs are more than twenty percent above their Reference Case levels (see also Chapter 2).

Table 4-2

ENERGY USAGE IN 2000:
THE LOW RENEWABLE ENERGY COSTS (LC) CASE
(Quadrillion Btu)

	Reference Case	LC Case	Percent Difference
Oil: Domestic	16.900	16.900	0.00
Imported	8.893	8.930	+0.42
Shale	0.730	0.730	0.00
Gas: Domestic	17.130	17.130	0.00
Imported	1.421	1.463	+2.96
Coal	36.613	36.673	+0.16
Nuclear	8.330	8.330	0.00
Renewables	10.765	10.765	0.00
Total Primary Energy	100.782	100.921	+0.14
Primary Inputs to Electric Generation	41.33	41.38	+0.12
Total Delivered Energy	69.15	69.26	+0.15
Percent Renewables in Primary Energy	10.7	10.7	
Percent Imports in Primary Energy	10.2	10.3	

Year 2000 costs for nonrenewable electric conversion and synthetic conversion facilities are also increased significantly (twenty to eighty percent) above their Reference Case levels (see Table 2-5). Thus, this case assumes that OPEC is able to increase prices for oil at a relatively rapid rate thorugh effective production limitations and other means. This case also assumes that recent rapid real cost increases for nonrenewable electric generation and synthetic fuel conversion facilities continue with only moderate attenuation for the 1980-2000 period.

Higher costs for nonrenewable energy supplies have a large, adverse impact on the growth and performance of the U.S. economy. Higher nonrenewables costs do not prevent continuing economic growth, but they do reduce the rate of growth. These economic effects appear as a result of lower growth in productivity and capital availability, and, consequently, lead to reduction in individual incomes and total economic production.

An overview of the energy and economic consequences of these higher costs is given in Table 4-3. The overall rate of growth is reduced by 0.06 percentage points so that, by 2000, real GNP is 1.23 percent, or $57 billion (1980$), less than GNP for the Reference Case in the year 2000. Annual productivity growth for the entire 1980-2000 period is reduced from 1.58 percent annually to 1.51 percent as a result of the higher costs. The annual rate of inflation is increased by 0.10 percentage points, reflecting the slower growth in productivity and reduced ability to absorb increases in input prices, especially higher energy prices. Finally, growth in energy use is reduced substantially, from 1.29 percent per annum to only 0.69 percent under the new higher conventional energy cost conditions. In fact, the decline in energy growth is much greater than the reduction in the growth of the economy, indicating a continuing improvement in the overall economic efficiency of energy use as compared to the Reference Case.

Higher costs for nonrenewable fuels, which provide most of our energy input, mean that any given volume of energy services now costs more to produce or purchase. Energy production now absorbs more of the scarce capital, labor, materials and other resources available in the economy. This additional claim on resources by the energy sector leaves less available for other uses. Thus, overall output potential declines (or increases less rapidly) as compared to the Reference Case. Since the same volume of economic input yields less total output, average productivity is reduced. Higher costs for nonrenewable energy sources also imply higher prices to producers and consumers purchasing fuels. In production, the price increases for energy raise overall

Table 4-3

OVERVIEW OF ENERGY AND THE ECONOMY:
THE HIGH NONRENEWABLE ENERGY COSTS (HC) CASE

	1980	1985	1990	1995	2000
Real GNP	2,630	2,992	3,419	3,908	4,557
Primary Energy Use	78.0	81.6	82.9	86.0	89.5
Energy-GNP Ratio	29.7	27.3	24.2	22.0	19.7

AVERAGE ANNUAL GROWTH RATES, 1980-2000

	Reference Case	HC Case
Real GNP	2.85	2.79
Primary Energy Use	1.29	0.69
Energy-GNP Ratio	-1.51	-2.03
Productivity	1.58	1.51
Inflation	7.24	7.34
Real Consumption	3.05	2.98
Real Investment	3.93	3.91
Capital-Labor Ratio	2.14	2.11

Units:
Real GNP in billions of 1980$.
Primary Energy in quadrillion Btu.
Energy-GNP Ratio in thousands of Btus per 1980$.
Growth Rates in percent per annum.
Productivity is real GNP per person employed.
Inflation is the rate of growth in the GNP price deflator.

production costs and, so, output prices. Consumers and producers respond by adjusting spending and input patterns away from energy and energy-intensive goods and services. But, these substitutions in production impose a cost since different inputs and outputs are not perfect substitutes. Further, the substitution of capital, labor and other inputs for energy involves a productivity cost. Greater inputs of capital and labor per unit of output mean that average productivity is reduced because economic output per unit of capital and labor input must be less. The result is slower growth of productivity, production and real incomes.

The energy reductions that accompany these changes do provide some compensating economic benefit. The energy efficiency of the economy is improved, reducing demands for energy. As energy demand is reduced, energy import requirements are curtailed. In turn, lower expenditures for imported energy result in exchange rate adjustments and in lower export volumes. Since fewer goods and services must be exported to pay for our oil and natural gas imports (recall that this analysis assumes that foreign trade is in approximate balance over the long term), more economic resources are now available for use in activities that sustain domestic consumption and investment. This partially compensates for the consequences of lower productivity arising from higher costs for nonrenewable energy.

As in the Reference Case, in the HC Case personal consumption expenditure is the dominant use of ouput. However, personal consumption expenditure absorbs most of the overall reduction in real GNP in the HC Case as compared to the Reference Case. This reduction in household spending is caused by the combination of price increases for energy and nonenergy goods and services and the income reduction that results from lower growth in productivity and productive capacity. Thus, consumers absorb much of the burden of adjustment to higher nonrenewable energy costs.

Business investment outlays are relatively unaffected by the changes in nonrenewable energy costs, accounting for less than four percent of the reduction in total real expenditure for the HC versus the Reference Case. The reasons for this are important in explaining why the economy is not more severely damaged by the higher costs for nonrenewable energy. There is a reduction in private savings as consumption outlays, in the face of higher prices, increase at a faster annual percentage rate than the percentage increase in personal income. But, there are compensating adjustments particularly involving the government sector. Current dollar revenues to governments rise with the value of economic activity, and this increase exceeds the rise in nominal government expenditure. The resulting decline in the overall government deficit moderates the impact of reduced savings on the value of investment. Further, the energy from new renewable and nonrenewable technologies becomes increasingly cost effective under the high-cost conditions. This results in a significant reduction in that portion of income required to subsidize new energy technologies. The net effect is that private investment can continue almost unchanged, despite the higher energy prices and reduced levels of private savings.

Higher prices for nonrenewable energy sources also have a substantial effect on relative prices and, from this, on demand and production patterns. The price of energy and energy-intensive goods rises relative to other goods and services. Thus, the rate of growth of energy prices is increased by 0.7 percentage points per annum (from 10.7 percent per year to 11.4 percent per year) while the annual increase in the aggregate price levels for all goods is increased by 0.07 percentage points, and services price growth goes up by only 0.04 points per year. Personal consumption patterns show a corresponding adjustment. Energy, in volume terms, declines relative to other purchases while services absorb a greater fraction of consumption. Thus, there is shift away from energy and energy-intensive puchases, like transportation, towards services. The slight shift from consumption to investment within overall final demand also leads to an increase in the relative importance of construction demand. Producers adjust to the new price conditions by altering their mix of inputs. The energy-intensity of production is reduced, with the place of this energy being taken by increased capital and labor input per unit of output. (By 2000, the overall structure of inputs to production involves 2.2 percent less energy per unit of output than in the Reference Case while capital intensity is 0.1 percent higher and labor intensity is up by 0.7 percent.)

Changes in industry growth rates reflect the adjustments in the volume and composition of final demand as well as the changes in input patterns. Growth rates in general are slowed. But industries catering to investment purchases are relatively less affected than those providing goods and services for personal consumption. Also, there is a shift away from energy and energy-intensive production. On balance, the principal shifts in industry growth are away from energy supply and toward communications, trade and service activities.

The more rapid rise in nonrenewable energy costs and in overall energy price, leads to a considerable slowing in the growth of energy use. This is compounded by the effects of less rapid growth in production and spending. Direct fuel purchases are cut, as spending and production become less energy-intensive. Table 4-4 shows the projected pattern of delivered energy consumption for 2000. Total use is reduced by almost 10 quadrillion Btu, or 14 percent relative to the Reference Case. Since the quantity of renewables is held constant, the energy reductions are distributed across conventional fuels. Oil and gas reductions are the largest, since these fuels experience more rapid price increases. Coal and electricity demand are somewhat less affected as the price increases for these fuels are more moderate. Lower electricity demand, in turn, implies fewer energy inputs to generation. Most of this cutback in electricity production is accommodated by reduced demand for coal by this sector, with a small reduction in oil and gas inputs. Demand reductions for oil and gas, coupled with a slight increase in domestic production (due to higher prices), leads to a substantial decline in energy imports. Crude oil imports are 61 percent less than in the Reference Case, while natural gas imports are 7 percent lower.

LOW RENEWABLE ENERGY COSTS AND HIGH NONRENEWABLE ENERGY COSTS (LHC) CASE

This Case combines the assumptions from the prior two scenarios discussed in this chapter. Thus, costs for renewable fuels and technologies are assumed to be significantly less than Reference Case levels, while costs for nonrenewable fuels and conversion technologies are significantly higher than their Reference Case levels.

Lower costs for renewable energy sources tend to moderate the adverse economic impacts that result from high nonrenewable costs alone. However, given the fact that nonrenewable energy sources still account for most of national energy use (still nearly 90 percent of all energy use in the year 2000), the net result of the combination of lower renewables and higher nonrenewable energy costs is a reduction in the growth of productivity and capacity as compared to the Reference Case. Lower rates of growth for productivity and productive capacity, in turn, lead to slower growth in output and incomes.

Table 4-5 provides a summary of the principal energy and economic effects from these combined changes in energy costs. Under the Reference Case cost conditions, real GNP growth averages 2.85 percent per annum over the 1980-2000 period. In the LHC Case, the corresponding figure is 2.80 percent annually. Thus, by 2000, real GNP is $46 billion (1980$), or 0.99 percent, less than for the Reference Case. Productivity growth

Table 4-4

ENERGY USAGE IN 2000:
THE HIGH NONRENEWABLE ENERGY COSTS (HC) CASE
(Quadrillion Btu)

	Reference Case	HC Case	Percent Difference
Oil: Domestic	16.900	16.900	0.00
Imported	8.893	3.496	-60.69
Shale	0.730	0.730	0.00
Gas: Domestic	17.130	13.972	18.44
Imported	1.421	1.315	-7.46
Coal	36.613	34.033	-7.05
Nuclear	8.330	8.330	0.00
Renewables	10.765	10.765	0.00
Total Primary Energy	100.782	89.541	-11.15
Primary Inputs to Electric Generation	41.33	39.43	-4.61
Total Delivered Energy	69.15	59.55	-13.88
Percent Renewables in Primary Energy	10.7	12.0	
Percent Imports in Primary Energy	10.2	5.4	

Table 4-5

OVERVIEW OF ENERGY AND THE ECONOMY:
THE LOW RENEWABLE ENERGY COSTS AND HIGH NONRENEWABLE ENERGY COSTS (LHC) CASE

	1980	1985	1990	1995	2000
Real GNP	2,630	2,993	3,421	3,914	4,568
Primary Energy Use	78.0	81.6	82.9	86.0	89.6
Energy-GNP Ratio	29.7	27.3	24.2	22.0	19.6

AVERAGE ANNUAL GROWTH RATES, 1980-2000

	Reference Case	LHC Case
Real GNP	2.85	2.80
Primary Energy Use	1.29	0.70
Energy-GNP Ratio	-1.51	-2.04
Productivity	1.58	1.52
Inflation	7.24	7.33
Real Consumption	3.05	2.98
Real Investment	3.93	3.96
Capital-Labor Ratio	2.14	2.12

Units:
Real GNP in billions of 1980$.
Primary Energy in quadrillion Btu.
Energy-GNP Ratio in thousands of Btus per 1980$.
Growth Rates in percent per annum.
Productivity is real GNP per person employed.
Inflation is the rate of growth in the GNP price deflator.

is reduced by 0.06 percentage points per annum and inflation is increased by 0.09 percentage points in comparison to the Reference Case projection. Also, the rate of growth in primary energy use is almost identical to that observed for the HC Case (High Nonrenewable Costs combined with Reference Case levels for renewable energy costs). However, the annual reduction in the energy-GNP ratio is slightly more rapid than for the HC Case (because energy use remains virtually unchanged while the GNP growth rate is slightly higher) and considerably more rapid than in the Reference Case (because average energy prices in the LHC Case are still much higher than for the Reference Case).

The mechanisms by which the lower renewables costs work to moderate the effects of higher costs for nonrenewables involve economic resources and investment. The cost reductions for renewable energy technologies and biomass fuels release capital, labor and other resources, allowing these resources to be used in other types of production. Equally important is the subsidy aspect of these new technologies. The reduction in renewables costs combined with the rise in conventional energy costs reduces the net cost disadvantage of renewables, thereby reducing the financial subsidies they require. This results in less diversion of investment funds away from private investment.[2] Relative to the Reference Case, the changes in the growth of capital, capital per worker, and capital prices are less under these cost conditions than those observed with high nonrenewable energy costs alone (the HC Case). Indeed, unlike the High Nonrenewable Energy Costs Case, real annual investment by the end of the century is actually higher than forecast for the Reference Case, although total capital stock has not recovered from earlier reductions in investment. As a consequence, the net adverse impacts on productivity, output, and incomes are smaller when low renewable energy costs occur in conjunction with increases in nonrenewable energy costs.

In terms of economic structure, this scenario is very much like the previous (HC) case, the difference being a matter of degree rather than direction. Price increases are larger for the outputs of the energy-intensive sectors than for the more services-oriented industries. The compositions of real final demand, real intermediate purchases, and real domestic output are

---

[2]We assume that these subsidies would be provided from public sources through tax incentives and other vehicles.

identical for this case and the HC Case. In terms of aggregate input patterns, the substitution of labor for capital and energy is not as pronounced as that occurring with only high nonrenewable energy costs, and the reduced energy intensity of production is also present, although to a lesser extent. These changes imply more capital per worker relative to the HC Case, but less capital per worker than exists in the Reference Case. Finally, the nature and magnitude of the reductions in national energy demand are comparable to, although somewhat smaller than, the case of higher nonrenewable energy costs alone. Thus, oil, gas and electricity use are all very slightly higher than for the HC Case but lower than for the Reference Case, while energy imports are lower for this case than the Reference Case.

HIGH RENEWABLES PENETRATION (HP) CASE

This scenario assumes that costs for nonrenewable and renewable fuels and energy conversion technologies remain at their Reference Case levels. However, for this scenario, market penetration (or utilization) for renewable technologies is about 40 percent higher than for the Reference Case by the end of the century. The distribution of this increase in utilization by technology is described in Chapter 2.

Increasing the market penetration (or use) of renewable energy supplies imposes a small cost on the economy in terms of slower growth in income and production. Table 4-6 summarizes the major characteristics of energy use and economic growth under conditions of higher market penetration by renewable energy technologies. Economic growth continues, but at a slightly slower rate than observed for the Reference Case. Annual growth in real GNP is reduced by 0.01 percentage points so, by 2000, the overall level of economic activity is lower by $13 billion (1980$) in this scenario compared to the Reference Case. Productivity growth declines by an average of 0.02 percentage points per year and the rate of inflation is increased marginally. The economy's use of energy is slightly higher. With slower economic growth, this implies a slight deterioration in the annual rate of reduction for the energy-GNP ratio. For the most part, the changes in economic variables do not differ markedly in this case from the Reference Case projection. However, the significant result for this case is the direction of the effects of greater renewables penetration. Greater use of renewable energy resources has negative economic impacts.

Table 4-6

OVERVIEW OF ENERGY AND THE ECONOMY:
THE HIGH RENEWABLES PENETRATION (HP) CASE

	1980	1985	1990	1995	2000
Real GNP	2,630	2,997	3,437	3,937	4,601
Primary Energy Use	78.0	83.7	88.3	94.0	101.0
Energy-GNP Ratio	29.7	27.9	25.7	23.9	22.0

AVERAGE ANNUAL GROWTH RATES, 1980-2000

	Reference Case	HP Case
Real GNP	2.85	2.84
Primary Energy Use	1.29	1.30
Energy-GNP Ratio	-1.51	-1.48
Productivity	1.58	1.56
Inflation	7.24	7.25
Real Consumption	3.05	3.04
Real Investment	3.93	3.90
Capital-Labor Ratio	2.14	2.12

Units:
Real GNP in billions of 1980$.
Primary Energy in quadrillion Btu.
Energy-GNP Ratio in thousands of Btus per 1980$.
Growth Rates in percent per annum.
Productivity is real GNP per person employed.
Inflation is the rate of growth in the GNP price deflator.

Higher penetration of renewables primarily affects the allocation of the economy's productive inputs and hence its productivity. The mix of renewable technologies in the Reference Case includes some that are cost-competitive in 1980, some that become cost-competitive over the period 1980 to 2000, and a few that provide no cost advantages through the end of the century. On balance, renewables require a greater commitment of economic inputs per unit of energy output than do the conventional energy sources they displace. Although, in the high penetration case, greater emphasis was placed on the relatively more cost-effective components of overall renewables supply, there still exists an overall cost disadvantage for renewables through the year 2000 as compared to nonrenewable resources. This results because more expensive centralized renewables technologies (such as OTEC and solar thermal "power towers") must be utilized to a greater degree in any high market penetration scenario than for a lower penetration scenario.

Higher penetration by renewables displaces conventional energy sources within the overall energy system. Accordingly, some of the capital, labor, and other inputs otherwise devoted to the supply of energy from conventional sources becomes available for use elsewhere. In addition, part of the reduction in use of nonrenewable fuels is reflected in lower petroleum and gas imports. These are accompanied by exchange rate adjustments leading to reductions in export volumes. Reduced production for export frees up economic inputs for use in sustaining domestic consumption and investment. However, because renewable energy resources involve a higher unit cost than the energy sources they displace, the inputs that are freed through these mechanisms are fully absorbed in supporting higher production levels for renewables. Further, the added cost from higher penetration requires additional inputs, beyond those released from conventional fuels. On balance, more economic inputs are used in supplying additional energy services to the economy from renewable resources than are released through lower use of nonrenewable supplies and reduced import requirements for crude oil and natural gas. Thus, the reallocation of energy demand from nonrenewable to renewable resources imposes a small loss in productivity, and, so, higher renewables penetration leads to slower growth in real output and incomes as compared to the Reference Case.

In addition, the energy changes affect investment which, in turn, affects future capital availability and productive potential. Slower economic growth leads to reduced government revenues (relative to expenditures) and to higher deficits. Even more important is the effect of changes in government financial support for new technologies. This support, whether in the form of direct public and private outlays or in the form of tax incentives and reduced tax revenue (increased tax expenditures), absorbs financial resources within the economy. This support is increased with the higher market penetration by renewable energy technologies. The net result of this additional financial support is to divert more private savings to government use and to the subsidy of new energy sources leaving less available for private investment. This is compounded by the effects of lower income and production growth and reduced rates of return, thereby slowing the growth of private saving and investment. The combined effect is for slower growth of investment and capital which reinforces the productivity effects of greater renewables penetration in retarding the overall rate of economic growth.

The impacts of higher renewables penetration on energy use are shown in Table 4-7. Direct uses of coal, petroleum, and gas are partially displaced by greater reliance on energy from dispersed renewables and electricity. The small increase in electricity demand occurs as structural changes in spending, input, and output patterns give rise to shifts in fuel use patterns, with electricity increasing at the expense of other delivered energy sources. In addition, the energy displacements and economic adjustments resulting from higher market penetration by renewable energy technologies alter relative energy prices. Since the relative price of electricity is reduced slightly, demand for this form of energy is further stimulated. The reductions in coal, oil and natural gas inputs to electric generation are, therefore, smaller than would result from a one-for-one displacement of conventional inputs to electric generation by renewables. Within primary energy, high renewables penetration has a large impact on petroleum and natural gas imports and somewhat less of an effect on overall coal consumption.

The higher level of total primary energy demand results from the shift toward electricity and the conversion losses associated with electricity. This shift toward greater reliance on electricity and the increase in primary energy inputs it entails more than offset the reduction in delivered energy requirements resulting from lower GNP in the HP Case compared to Reference Case. The net result is the 0.26 percent increase in total primary energy use noted in Table 4-7.

Table 4-7

ENERGY USAGE IN 2000:
THE HIGH RENEWABLES PENETRATION (HP) CASE
(Quadrillion Btu)

	Reference Case	HP Case	Percent Difference
Oil: Domestic	16.900	16.900	0.00
Imported	8.893	7.651	-13.97
Shale	0.730	0.730	0.00
Gas: Domestic	17.130	17.130	0.00
Imported	1.421	0.291	-79.52
Coal	36.613	34.925	-4.61
Nuclear	8.330	8.330	0.00
Renewables	10.765	15.085	+40.13
Total Primary Energy	100.782	101.042	+0.26
Primary Inputs to Electric Generation	41.33	41.96	+1.52
Total Delivered Energy	69.15	69.02	-0.18
Percent Renewables in Primary Energy	10.7	14.9	
Percent Imports in Primary Energy	10.2	7.9	

HIGH RENEWABLES PENETRATION AND LOW RENEWABLES COSTS (HPLC) CASE

This case combines the energy cost and market penetration assumptions of the Low Renewables Cost Case (LC) and the High Renewables Penetration Case (HP). Thus, renewable technology costs (per unit capacity) in the year 2000 are ten to forty percent below their Reference Case levels (and the rate of increase for biomass fuels is also reduced from Reference Case levels), while the market penetration of renewables is increased about 40 percent above that for the Reference Case.

The simultaneous expansion of renewable supplies and reductions in renewables costs are virtually neutral in terms of their overall effects on production and incomes. These results are summarized in Table 4-8. Compared to the Reference Case, real GNP is slightly lower through the mid-1990s. After 1996, however, the energy changes inherent in the HPLC Case exert a positive influence on output and incomes. By 2000, real GNP is $5.5 billion (1980$), or 0.12 percent, higher as a consequence of the combined effects of lower costs and higher market penetration for renewable energy technologies.

The mechanisms underlying this approximate cancelling out of effects are related to real resource allocation and investment changes. Higher market penetration of renewable energy technologies displaces some energy imports and domestic production of nonrenewable energy resources. This permits the release of capital, labor, and other inputs from production for export and for the supply of conventional energy to other uses within the economy. At the same time, however, additional use of renewable energy technologies places claims on capital, labor and other inputs. The net of these two effects determines the nature of the productivity effects. In this scenario, which assumes cost reduction and quantity increase for renewable energy technologies, renewables require a slightly smaller withdrawal of the available inputs within the economy than are released by reduced demands for domestic and imported nonrenewable energy resources. The remaining inputs are available for other uses to produce goods and services for domestic consumption and investment. On balance, therefore, the real resource effects of higher renewables penetration and lower renewables costs are very slightly favorable to productivity growth, output, and income as compared to the Reference Case.

The overall economic impact of greater use of and lower costs for renewable energy technologies and

Table 4-8

OVERVIEW OF ENERGY AND THE ECONOMY:
THE HIGH RENEWABLES PENETRATION AND LOW RENEWABLES COSTS (HPLC) CASE

	1980	1985	1990	1995	2000
Real GNP	2,630	2,998	3,441	3,946	4,619
Primary Energy Use	78.0	83.7	88.4	94.1	101.2
Energy-GNP Ratio	29.7	27.9	25.7	23.8	21.9

AVERAGE ANNUAL GROWTH RATES, 1980-2000

	Reference Case	HPLC Case
Real GNP	2.85	2.86
Primary Energy Use	1.29	1.31
Energy-GNP Ratio	-1.51	-1.50
Productivity	1.58	1.59
Inflation	7.24	7.24
Real Consumption	3.05	3.06
Real Investment	3.93	3.96
Capital-Labor Ratio	2.14	2.14

Units:
Real GNP in billions of 1980$.
Primary Energy in quadrillion Btu.
Energy-GNP Ratio in thousands of Btus per 1980$.
Growth Rates in percent per annum.
Productivity is real GNP per person employed.
Inflation is the rate of growth in the GNP price deflator.

resources is also affected by certain temporal trends in investment. Until the early 1990s, the net effect of greater use of but lower cost for renewables is to require a larger subsidy to support these activities. This subsidy absorbs some of the financial resources that would have been available for private investment. This causes a slight slowing in investment and capital growth. After the early 1990s, the cost reductions for renewables dominate, so that renewables require a smaller subsidy. This permits more savings to be channeled into private investment, accelerating the growth of investment. By 2000, the economy's stock of productive capital has recovered and, in fact, is slightly higher than in the Reference Case.

There are few substantive differences between the structures of final demand, aggregate and intermediate input patterns, and domestic production under energy conditions prevailing in the HPLC Case and those in the Reference Case. The only real difference concerns the growth in capital per worker over the period 1980 to 2000. In line with the investment changes noted previously, growth in capital per worker is slower through the middle 1990s and accelerates slightly thereafter relative to the Reference Case.

The changes in energy use patterns are similar to those of the High Renewable Penetation (HP) Case. Electricity and dispersed renewables displace some coal, oil, and gas in delivered energy consumption. By the year 2000 the slight gain in economic perfornance leads to an increase in demand for electricity above the levels prevailing for the Reference Case. The increased reliance on renewable resources and electricity involves higher primary energy consumption in spite of the moderate reduction in coal use and the significant displacements of petroleum and natural gas imports.

HIGH RENEWABLES PENETRATION AND HIGH NONRENEWABLES COSTS (HPHC) CASE

This scenario combines the high renewables penetration assumptions from the HP Case and the high nonrenewable energy costs assumptions from the HC Case. Thus, renewables supply about 15 quads of energy in 2000 (as compared to less than 11 quads in the Reference Case). It should be noted that, because some renewables costs must be subsidized even with high nonrenewable costs prevailing, there is no necessary causal link between high nonrenewable costs and high renewables penetration rates (as both are defined for this study).

Even with the costs of nonrenewable energy resources (both fuels and electric/synfuel conversion technologies) increased substantially above their already high Reference Case levels, the average costs of energy from renewable resources are still higher than for energy from nonrenewable resources through the 1990s. For this reason, of all the scenarios considered in this analysis, this combination of energy cost and market penetration changes has the largest adverse impact on the growth and performance of the economy. A summary of the energy and economic consequences from these changes appears in Table 4-9. The overall rate of economic growth is reduced by 0.07 percentage points per year. Relative to the Reference Case, real GNP is lower by $28 billion in 1990 and $59 billion by the end of the century (1980$). Annual productivity growth is reduced by 0.07 percentage points, and the annual rate of inflation is increased by 0.10 points. Growth in energy use is reduced significantly, from 1.20 percent annually for the Reference Case to only 0.76 percent under these new energy conditions.

These effects result from a combination of adjustments to high market penetration by renewable energy technologies and high costs for nonrenewable energy resources. Both of these changes involve economic costs in the form of productivity and income losses. Higher costs for nonrenewables involve a direct economic cost by absorbing greater amounts of capital, labor and other inputs in the production and purchase of energy supplies and an indirect cost by inducing the substitution of capital and labor for energy input. These substitutions result in gross productivity losses. Higher penetration by renewable energy technologies involves similar costs. In particular, as additional economic resources are absorbed in the energy sector and as the shift from lower cost conventional fuels to higher cost renewable supplies takes place, economic resources are diverted toward energy supply activities leaving less available for other uses. Thus, overall productivity and output growth rates decline (i.e., over time, they increase less rapidly) relative to the Reference Case).

The price increases for energy incorporated into this scenario affect production costs and, in turn, output prices. Consumers and producers respond by adjusting spending and input patterns, primarily away from energy and energy-intensive goods and services. But, substitution possibilities are limited by the availability of capital and labor especially since domestic energy production now requires more of these inputs. The restructuring of spending and input patterns imposes a further productivity cost since inputs and outputs are not perfect substitutes in production and consumption.

Table 4-9

OVERVIEW OF ENERGY AND THE ECONOMY:
THE HIGH RENEWABLES PENETRATION AND HIGH NONRENEWABLES COSTS (HPHC) CASE

	1980	1985	1990	1995	2000
Real GNP	2,630	2,990	3,415	3,902	4,554
Primary Energy Use	78.0	81.6	82.9	86.0	90.6
Energy-GNP Ratio	29.7	27.3	24.3	22.0	19.9

AVERAGE ANNUAL GROWTH RATES, 1980-2000

	Reference Case	HPHC Case
Real GNP	2.85	2.78
Primary Energy Use	1.29	0.76
Energy-GNP Ratio	-1.51	-1.97
Productivity	1.58	1.51
Inflation	7.24	7.34
Real Consumption	3.05	2.98
Real Investment	3.93	3.91
Capital-Labor Ratio	2.14	2.09

Units:
Real GNP in billions of 1980$.
Primary Energy in quadrillion Btu.
Energy-GNP Ratio in thousands of Btus per 1980$.
Growth Rates in percent per annum.
Productivity is real GNP per person employed.
Inflation is the rate of growth in the GNP price deflator.

Investment and growth in capital stocks are also affected by these energy changes. Household expenditures increase due to higher prevailing price levels for goods and services. The result is reduced household saving. Government revenues rise as a result of the more rapid rate of inflation. This tends to lower the deficit. However, the greater penetration of renewables involves additional subsidies for these technologies, which increases the deficit (through revenue loses and direct subsidies). Thus, not only are private savings reduced but also the government deficit rises, diverting more financial resources away from private investment. The slower growth of investment translates into slower capital growth, accentuating the effects of reductions in productivity growth on GNP growth.

The increases in nonrenewable energy costs and in the market penetration of renewable technologies have the direct effect of raising the price of energy. Indirectly, these energy changes lead to higher prices for capital inputs by reducing capital availability and the supply of investment funds. These cost impacts are largest for energy and capital intensive industries, such as the manufacturing sector, and are less for other industrial sectors, such as services. The sectoral mix of final demand is altered in respose to these price changes -- energy purchases are cut, manufacturing and transportation purchases are somewhat reduced, while services purchases are increased. Input patterns are shifted away from energy and, to a lesser extent, from capital. Labor input increases in relative importance (which is beneficial for employment but detrimental for the average rate of productivity increase).

Energy use from nonrenewables is displaced through both the adjustments to higher costs (which results in reduced overall demand for energy) and the increased utilization of renewable supplies. Table 4-10 compares energy consumption patterns for the year 2000 for this scenario with the Reference Case. Total use of delivered energy for this scenario is reduced by more than 9 quads, or 13 percent, from Reference Case levels. Since the quantity of renewable energy used is higher, the energy reductions among conventional inputs are sustantial. Reductions in the use of oil and natural gas are the largest, since more rapid price increases are assumed for these fuels. Coal use is reduced to a lesser extent and electricity demand experiences the smallest reduction. Slightly lower electricity demand in conjunction with the increase in renewables capacity implies substantial reductions in coal, oil, and gas inputs to the electric sector. Finally, in response to price-induced demand changes and

Table 4-10

ENERGY USAGE IN 2000:
THE HIGH RENEWABLES PENETRATION AND HIGH NONRENEWABLES COSTS (HPHC) CASE
(Quadrillion Btu)

	Reference Case	HPHC Case	Percent Difference
Oil: Domestic	16.900	16.900	0.00
Imported	8.893	2.374	-73.30
Shale	0.730	0.730	0.00
Gas: Domestic	17.130	13.280	-22.48
Imported	1.421	1.256	-11.61
Coal	36.613	32.676	-10.75
Nuclear	8.330	8.330	0.00
Renewables	10.765	15.085	+40.13
Total Primary Energy	100.782	90.631	-10.07
Primary Inputs to Electric Generation	41.33	40.52	-1.96
Total Delivered Energy	69.15	59.91	-13.36
Percent Renewables in Primary Energy	10.7	16.6	
Percent Imports in Primary Energy	10.2	4.0	

displacements of nonrenewable fuels by renewable energy resources, petroleum and natural gas imports decline by more than 60 percent and domestic natural gas production declines significantly. Use of coal also is diminished, and total primary energy consumption ultimately is reduced by 10 percent from Reference Case levels.

HIGH RENEWABLES PENETRATION, LOW RENEWABLES COSTS, AND HIGH NONRENEWABLES COSTS (HPLHC) CASE

This case combines all three sets of energy cost/penetration changes from the Reference Case. It assumes that renewables penetration rates are high (40 percent above the Reference Case), nonrenewable fuel and electric/synfuel technology costs are significantly higher than the Reference Case, and renewable technology costs are ten to forty percent lower than the Reference Case.

When high nonrenewable energy costs occur simultaneously with higher penetration of and lower costs for renewable energy technologies, the economy again is affected adversely. High nonrenewable energy costs result in damage regardless of the prevailing conditions for renewable supplies. The principal features of this scenario are reported in Table 4-11. Virtually all aspects of the economy are made worse by this combination of energy changes. In comparison to the Reference Case, economic growth and growth in productivity are reduced by 0.05 percentage points per year and inflation is increased by 0.09 percentage points per year. Growth in energy use is slowed and the average efficiency of energy use is increased although these effects do not come close to offsetting the economic costs associated with the HPLHC Case.

The damage to the economy under these conditions is less than that experienced with either high nonrenewable energy costs alone (HC) or high nonrenewable energy costs combined with high penetration (HPHC) of renewable energy technologies (and Reference Case levels of renewable costs). However, the damage is greater than that observed under conditions of the LHC Case (i.e., Reference Case levels of utilization of renewable technologies but lower costs for renewables and higher costs for nonrenewables). These results are related to real resource effects and changes in investment patterns. When nonrenewable costs are increased, overall energy production requires a larger commitment of the economy's productive inputs. For renewable energy technologies, the increase in renewables utilization claims more economic inputs while the cut in renewables costs reduces the input requirement. The net effect for renewables is the release of a small volume

Table 4-11

OVERVIEW OF ENERGY AND THE ECONOMY:
THE HIGH RENEWABLES PENETRATION, LOW RENEWABLES COSTS
AND HIGH NONRENEWABLES COSTS (HPLHC) CASE

	1980	1985	1990	1995	2000
Real GNP	2,630	2,992	3,418	3,910	4,570
Primary Energy Use	78.0	81.6	82.9	86.1	90.7
Energy-GNP Ratio	29.7	28.0	24.3	22.0	19.8

AVERAGE ANNUAL GROWTH RATES, 1980-2000

	Reference Case	HPLHC Case
Real GNP	2.85	2.80
Primary Energy Use	1.29	0.76
Energy-GNP Ratio	-1.51	-1.99
Productivity	1.58	1.53
Inflation	7.24	7.33
Real Consumption	3.05	2.99
Real Investment	3.93	3.96
Capital-Labor Ratio	2.14	2.12

Units:
Real GNP in billions of 1980$.
Primary Energy in quadrillion Btu.
Energy-GNP Ratio in thousands of Btus per 1980$.
Growth Rates in percent per annum.
Productivity is real GNP per person employed.
Inflation is the rate of growth in the GNP price deflator.

of economic resources. This helps to limit the damage associated with higher costs for conventional fuels. However, higher penetration still imposes an economic cost.

Higher costs for nonrenewable energy result in increases in general price levels and decreases in real income. These changes, in turn, lead to reductions in private savings, thereby reducing investment. Increased prices for nonrenewable energy supplies also permit reductions in the subsidy required for renewable energy supplies, releasing these funds for private investment. Lowering renewables costs further reduces the subsidy. The ultimate result of these financial effects is to moderate somewhat the negative investment impacts associated with the higher prices for conventional fuels, higher costs for fossil and nuclear energy conversion technologies, and greater utilizaton of renewable resources.

The structure of the economy in this scenario is very similar to that in the previous case (the HPHC Case). Prices of energy inputs and of energy and energy-intensive outputs are increased relative to prices for other goods and services, such as labor inputs or services sector outputs. There is a small shift in the composition of final demand away from consumption and toward investment. Also, the purchase pattern of final demand shifts slightly toward construction and equipment (associated with the investment effect) and services (associated with the price changes), and away from energy. Inputs to production become less energy-intensive and more capital- and labor-intensive. One consequence of these input shifts is a slowing in the rate of growth of productivity. Finally, energy use declines as both final and intermediate energy demand is cut back. Low costs for renewable energy sources by themselves would lead to small increases in total energy consumption. However, because high costs for nonrenewable energy resources result in substantial reduction in energy use, the net effect is a considerable slowing in the growth of energy demand as compared to the Reference Case. The reductions in demand are concentrated in use of oil and natural gas, thereby producing large percentage reductions in energy imports.

5 Energy Changes and Economic Performance

The previous chapter examined seven energy/economic scenarios developed for this analysis. Each of the scenarios was compared to the Reference Case. In this way we established both the nature and the magnitude of the effects of energy changes on the growth and structure of the nation's economy. This chapter draws broad comparisons across all the scenarios to provide a perspective on the economic effects of energy costs and supply changes.

GENERAL ECONOMIC AND ENERGY EFFECTS

Macroeconomic performance in the eight different energy cases is summarized in Table 5-1. This table contains information on the consequences of higher market penetration by renewable energy technologies, of lower costs for renewable energy and of higher costs for nonrenewable energy. From this table it can be seen that, compared to the Reference Case,

- real GNP growth and future levels of production and incomes are reduced in all of the energy scenarios except those which incorporated reductions in renewable energy costs and the Baseline Trend for nonrenewable energy costs;
- productivity growth is slowed in most of the scenarios; and
- inflation is accelerated in most of the scenarios.

In all of the energy scenarios, except the one where a reduction in renewables costs was unaccompanied by changes in other energy conditions (i.e., increases in

Table 5-1

OVERVIEW OF MACROECONOMIC EFFECTS

Case	Characteristics of Each Case			Cumulative Real GNP Effect	Average Annual Real GNP Growth	Average Annual Productivity Growth	Average Annual GNP Inflation Rate
	Renewables Penetration	Renewables Costs	Nonrenewables Costs				
Reference				0	2.85	1.58	7.24
LC		L		89	2.87	1.59	7.23
HPLC	H	L		-4	2.86	1.59	7.24
HP	H			-122	2.84	1.56	7.25
LHC		L	H	-431	2.80	1.52	7.33
HPLHC	H	L	H	-470	2.80	1.53	7.33
HC			H	-509	2.79	1.51	7.34
HPHC	H		H	-577	2.78	1.51	7.34

Notes:

1. Notation used in the specification of the cases is:
   Renewables Penetration is High = H (otherwise Reference Case penetation)
   Renewables Costs are Low = L (otherwise Reference Case costs)
   Nonrenewables Costs are High = H (otherwise Reference Case costs)

2. Cumulative real GNP effect is the sum of the difference of real GNP from the Reference Case, 1980 to 2000, in billions of 1980 dollars.

3. Average real GNP growth, productivity growth and inflation rates are in annual percentages for the 1980 to 2000 period.

renewables penetration or nonrenewables costs),[1] adverse macroeconomic effects were observed -- slower growth of production and real incomes, slower growth of productivity, more rapid inflation, and/or cumulative GNP losses. The reasons for the adverse effects are analyzed in Chapter 4, but, in essence, they involve the effect of absorbing more economic resources for energy supply (either through higher costs for nonrenewable fuels or through greater market penetration of relatively high cost renewables) leaving less economic resources available for the production of other goods and services.

Given that these macroeconomic effects are negative or adverse, an important concern is their magnitude and significance. In relative terms, the effects are not large. The maximum relative reduction in average annual rates of real economic growth is only about two percent (or 0.1 percentage points in absolute terms), but even in the slowest growth case, economic growth continues at about 2.8 percent annually. The productivity growth reduction is of similar absolute magnitude (about 0.1 percentage points) but annual productivity growth of 1.5 percent remains. Finally, in the worst case average annual inflation rates are increased only from 7.2 percent to 7.3 percent, a relatively small increment. Thus, for each of these indicators, the impact of the energy changes is noticeable but relatively small from a macroeconomic perspective. The general growth path of economic variables is not greatly altered. In this sense, the macroeconomic effects of the energy changes, although adverse, can be absorbed without much effect on future increases in production, real income and purchasing power.

However, even though these economic impacts can be sustained without major economic change, this does not alter the fact that these costs exist and that they can be significant in absolute magnitude. The reduction in the average annual rate of economic growth from 2.85 percent to 2.78 percent corresponds to a cumulative loss in production and income of $577 billion, in constant 1980 dollars. This loss is equivalent to a one-time, lump-sum tax of around $7,500 for each household in the United States in 1980. It should be stressed that this loss does not occur as an absolute reduction in GNP, but only as a reduction in the level of GNP from what it would have become. Thus, real incomes actually increase

---

[1]Note that the High Renewables Penetration Low Renewables Costs (HPLC) Case had approximately neutral macroeconomic effects compared to the Reference Case.

over time in all scenarios analyzed here. However, these cumulative reductions still represent costs in the sense of potential losses of income and production resulting directly from the energy changes inherent in the different Cases. Thus, while the losses due to the energy changes can be sustained, they are quantitatively significant and do represent a real burden on everyone in the economy. Furthermore, the significance of these losses increases if they are related to the magnitude of the change in the energy system rather than expressed relative to the entire economy (see below).

One further set of conclusions suggested by the information on macroeconomic effects concerns the different impacts of the different types of energy changes. Table 5-1 shows a systematic difference between the economic effects of the first four energy changes and those of the last four changes; the magnitude of the economic costs of the last four energy changes is substantially greater. The one feature common to the last four changes is the higher costs of nonrenewable energy. This suggests that changes in costs or prices for conventional, nonrenewable fuels can have a much greater economic effect than changes in costs or market penetration of renewable energy forms. The reason for the marked difference in effects stems from the sheer difference in size of the markets for these various types of energy. Nonrenewable fuels comprise the major part of the energy demand (well over 80 percent of primary energy use in all Cases analyzed). Therefore, each percentage point change in unit costs of energy from these fuels is of far greater economic consequence than a percent change in the cost of energy from renewables. It follows that the economic damage done by an increase in the cost of nonrenewable energy is considerably greater than for an equi-proportional increase in the cost of renewable energy. Similarly, a reduction in costs of nonrenewables has a far greater economic benefit than a reduction in costs of renewable energy forms.

An overview of growth in energy usage in the eight different energy scenarios is give in Table 5-2. This table shows U.S. primary energy input, its growth and its relationship to real GNP. Total energy use grows under all eight scenarios, from around 78 quadrillion Btu in 1980 to 90 to 100 quads in 2000. These figures correspond to average growth rates of from 0.7 to 1.3 percent annually. In all cases, this growth rate is lower than the growth rate for economic output in the economy, i.e., for all eight scenarios the energy-GNP ratio is declining over time as the economy becomes less energy intensive.

Table 5-2

OVERVIEW OF ENERGY EFFECTS

Case	Characteristics of Each Case			Primary Energy Use (Year 2000)	Average Annual Growth in Energy Use	Average Annual Change in Energy-GNP Ratio
	Renewables Penetration	Renewables Costs	Nonrenewables Costs			
Reference				100.8	1.29	-1.51
LC		L		100.9	1.30	-1.52
HPLC	H	L		101.2	1.31	-1.50
HP	H			101.0	1.30	-1.48
LHC		L	H	89.6	0.70	-2.04
HPLHC	H	L	H	90.7	0.76	-1.99
HC			H	89.5	0.69	-2.03
HPHC	H		H	90.6	0.76	-1.97

Notes:

1. Notation used in the specification of the cases is:
   Renewables Penetration is High = H (otherwise Reference Case penetation)
   Renewables Costs are Low = L (otherwise Reference Case costs)
   Nonrenewables Costs are High = H (otherwise Reference Case costs)

2. Primary energy input is quadrillion Btu in 2000.

3. Average energy growth is average annual percentage growth rate in primary energy input over the 1980 to 2000 period.

4. Change in the primary energy-real GNP ratio is average rate of change for 1980 to 2000.

In contrast to the relatively small economic changes resulting from different energy conditions, there are major impacts on overall energy use across the different scenarios. By 2000, primary energy input varies to as much as 10 percent below the Reference Case for those Cases which incorporate higher nonrenewable energy costs. Growth rates for energy use are, in some case, halved relative to the Reference Case. For those cases where nonrenewable energy costs are the same as the Reference Case (i.e., the first four cases in Tables 5-1 and 5-2), primary energy use in 2000 is virtually identical to primary energy use in the Reference Case. In those cases where nonrenewable costs are higher than for the Reference Case (i.e., the latter four cases in Tables 5-1 and 5-2), primary energy use in the year 2000 is about 10 percent below that in the Reference Case. The resulting nonrenewable energy cost increases for these cases cause a marked slowing in the growth of energy demand. Growth in demand for primary energy is reduced to 0.7 percent annually, compared to 1.3 percent in the Reference Case, and the energy intensity of the economy falls by 2 percent annually, rather than the 1.5 percent rate of decline in the Reference Case. In contrast, by themselves, changes in costs or penetration of renewable energy forms have almost no effect on total energy demand. This is principally a result of the small relative role played by renewables in total energy demand and the fact that the overall effective price of energy is not greatly affected by changes in renewables costs.

ENERGY-ECONOMY RELATIONSHIPS[2]

The broad economic and energy adjustments to the changes in energy cost and supply conditions show a consistent pattern in that the economic changes are proportionately less than the energy changes. For example, the shift from the Reference Case to conditions of high costs for nonrenewable energy (HC) involves primary energy input in 2000 that is 10.1 percent lower whereas real GNP is only 1.2 percent lower. The ratio of economic change to energy change is only 0.12. This suggests a considerable degree of flexibility in the

---

[2]This section focuses on the economic impacts of higher nonrenewable energy costs (as compared to the Reference Case) because higher nonrenewables costs has a much larger economic and energy impact than either higher penetration for renewables or lower costs for renewables (see also Table 5-1).

economy that permits an adjustment to higher energy prices or lower energy quantities to be achieved without a corresponding reduction in the level of output or production from the economy. This flexibility works to reduce, although it does not eliminate, the economic effects (or costs) associated with a rise in energy costs. This section summarizes the adjustment mechanisms that provide this energy-economy flexibility.

The initial cause for economic change is a change in energy costs. This change in energy costs may be the result of one or more of three factors:

- higher fuel and capital costs for conventional (fossil and nuclear) energy forms;
- lower fuel and capital costs for energy from renewable sources; and/or
- higher market penetration by (relatively expensive)[3] renewable energy sources.

But, regardless of the scenario, the central change incorporated into that scenario for this study is (at least implicitly) a change in prices of energy to intermediate producers or final purchasers.

For the HC Case, for example, the energy cost increases result from an increase in the costs of nonrenewable fuels and of certain electric and synthetic fuel conversion technologies. Table 5-3 shows the changes in input and output prices due to the higher energy costs. On the input side, capital and materials inputs both show a small rise in price relative to labor (this relative price increase reflects the fact that capital and materials have a higher "energy content" than does labor) whereas energy inputs show a large increase in relative price. Output prices also change. These changes reflect the differing energy content (and capital, labor and materials content) of the different types of output. Energy rises in relative price by over 13 percent whereas agriculture, construction, manufacturing and transporation show a 0.4 percent decline in relative price and services prices decline by 0.6 percent relative to the average. In both input and output markets, however, the dominant change is the rise in energy prices relative to all other goods and services.

Consumers (accounting for 65 percent of all final demand purchases) and other purchasers of finished goods and services respond to the rise in energy prices by shifting their purchase patterns away from energy. With the increase in nonrenewable energy costs in the HC Case

---

[3]As compared to nonrenewable energy resources.

Table 5-3

CHANGES IN INPUT AND OUTPUT PRICES
(Price Rises in HC Case Relative to the Reference Case in 2000, in percent)

Final Output Prices	Absolute Change in Price	Change Relative to Average Price
Agriculture, Construction	1.1	-0.3
Manufacturing	1.0	-0.4
Transportation	1.0	-0.4
Services	0.8	-0.6
Energy	15.0	13.4
Average	1.4	0

Input Prices	Change Relative to Labor Price
Capital	1.6
Labor	0
Energy	11.4
Materials	1.0

relative to the Reference Case, people still buy fuels and still use energy services. Also, the level of demand of energy services still grows over time but less rapidly than otherwise. In a relative sense, energy and energy-intensive goods play a smaller part in final demand over time, while services and other non-energy goods and services increase in importance. The motivation is simply the objective of purchasers to obtain the highest standards of living in the face of a new price structure in which energy is relatively more expensive than before. The result is a reduction in the energy content of final purchases of goods and services. Table 5-4 gives a quantitative summary of these effects -- the volumes of energy and transportation purchases, relative to total final demand, are lower while the volumes of other purchases are all increased as a result of higher energy prices.

Producers also respond to the rise in energy prices by shifting purchases away from energy toward other inputs. This takes the form of gradual changes and adjustments to growth rates rather than sudden shifts. However, the effect is to reduce the relative role of energy and energy-intensive inputs and, correspondingly, to increase the role of labor and capital inputs. The largest adjustment involves energy input but there are also noticeable changes in other factors of production. Table 5-5 shows the change in aggregate input patterns caused by the higher energy prices. The energy intensity of production is reduced by over 2 percent for the HC Case relative to the Reference Case while capital and, in particular, labor are used in place of this energy.

Final demand changes are adjustments in what the economy produces. The input pattern changes are adjustments in how this production takes place. Thus, substitution away from energy use is apparent in both what is produced and the manner in which it is produced. These shifts are mutually reinforcing. The result is a significant reduction in the energy intensity of the economy. The sequence starts with the costs of energy supply and with energy prices. With price increases acting on the purchase decisions of consumers and producers, the role of energy in the econony is reduced. This, in essence, is the source and nature of the flexibility in the economy that permits the energy-intensity (as indicated, for example, by the energy-GNP ratio) to fall in response to higher energy prices. The shift away from energy also means that a rise in energy prices causes a reduction in energy quantity demanded that exceeds any reduction in economic activity, or real GNP. Thus, the proportionate economic reduction is less than the proportionate reduction in energy use.

Table 5-4

CHANGES IN THE COMPOSITION OF FINAL DEMAND AS A RESULT OF HIGHER NONRENEWABLE ENERGY COSTS (HC CASE)
(Percent of Total Real Final Demand in 2000)

	Reference Case	Higher Nonrenewable Energy Prices (HC) Case
Agriculture, Construction	12.9	13.0
Manufacturing	34.9	35.0
Transportation	3.7	3.6
Services	45.7	45.8
Energy	2.8	2.6
Total	100.0	100.0

Table 5-5

CHANGES IN THE PATTERN OF INPUTS TO PRODUCTION AS A RESULT OF HIGHER NONRENEWABLE ENERGY COSTS (HC CASE)
(Share of Real Input in 2000)

	Reference Case	Higher Nonrenewable Energy Prices (HC) Case	Percent Change
Capital	0.1595	0.1597	0.13
Labor	0.2061	0.2076	0.73
Energy	0.0593	0.0580	-2.19
Materials	0.5751	0.5747	-0.07
Total	1.0000	1.0000	0.00

Although this flexibility in the economy permits the economic effects of energy price increases to be contained, it does not eliminate them. A rise in energy prices or reduction in energy quantities will still involve some net cost to the economy. There are several reasons for this cost. Any energy cost rise that absorbs more economic resources (capital, labor and intermediate materials) prevents these resources being used for other production, thereby lowering the overall output potential of the economy. Also, the substitution process in production, whether substitution between different inputs within a particular type of production process or substitution of resources among different types of production, involves some cost both in making the shifts and also because inputs are not perfect substitutes. Further, the trend toward a lower energy intensity is achieved through a shift toward higher capital and labor intensity of production. In Table 5-5, the capital plus labor input per unit of output rises from 0.3656 in the Reference Case to 0.3674 under higher energy prices (the HC Case). The average output per unit of capital plus labor input therefore declines by 0.46 percent, and average labor productivity falls by 0.72 percent. In short, use of more capital and labor per unit of output means less output per unit of these inputs, that is, a decline in productivity. Thus, while the average energy productivity of the economy rises, the average productivity of capital and labor falls. Since the quantities of these inputs available to the economy are limited, a decline in productivity is likely to translate into a decline in production, or real GNP. (However, the increased requirement for labor input may also operate to raise the level of employment, or reduce unemployment. In such a situation, only part of the productivity decline would carry over into a reduction in the real GNP.)

The conclusion from this analysis is that there is a considerable degree of flexibility in the economy, in terms of what is produced and how it is produced, and that this flexibility permits a large fraction of any energy change to be accommodated without a reduction in production and incomes. However, there will remain some economic cost, in terms of slower growth of real incomes, associated with any rise in energy costs. These conclusions account for two of the general features in the eight scenarios analyzed for this study -- (1) higher energy costs cause some economic penalty, in the form of slower growth, and (2) these economic changes are much smaller, in relative terms, than the associated energy changes.

ECONOMIC IMPACTS OF ENERGY CHANGES

Each of the three energy changes analyzed in this study -- increased penetration of renewables, decreased costs of renewables, and increased costs of nonrenewable energy forms -- has an effect on incomes and production in the economy. To assess the relative directions and magnitudes of these effects, total real GNP over the 1980 to 2000 period is used as an indicator. The change in cumulative or total real GNP is given in Table 5-6 for each of three types of energy changes. This section analyzes the economic effects of each type of energy change and also quantifies the range of economic impacts associated with each energy change.

The first section of Table 5-6 shows the effect of each type of energy change in isolation. These effects are the averages of the impacts, across all combinations of specifications of the remaining energy conditions. They provide a summary indicator of the economic impact of each type of energy change. The implications of these impact figures are:

- reduced costs for renewable energy have a beneficial effect on economic performance;
- increased penetration or use of renewable energy above the level specified in the Reference Case has an adverse effect on the economy;
- increased costs for nonrenewable energy have an adverse economic effect;
- none of these effects is particularly large relative to the economy as a whole, although each impact is significant in absolute magnitude;
- the effect of increased costs for nonrenewable energy is considerably larger than the effect of changes in cost and/or quantity conditions for renewable energy forms.

These results show that higher costs for energy supply, whether of conventional or of renewable energy forms, have adverse economic effects and that more widespread use of renewable forms of energy also has an adverse impact on the economy. The reasons for these results all relate to the quantity of economic resources required to provide each unit (e.g., a million Btus) of energy. If more input of capital, labor and intermediate materials is needed per unit of energy service output, then less of these economic resources are available for use elsewhere. Consequently, production of other goods and services is less, and the total output potential of the economy is reduced.

Table 5-6

CHANGE IN CUMULATIVE REAL GNP, 1980-2000, DUE TO ENERGY CHANGES
(Figures in Billions of 1980 Dollars)

		Change in Cumulative GNP as Percentage of Reference Case GNP in the Year[a]	
		1980	2000
Average Effects[b]			
Reduced Renewable Energy Costs	98	3.7	2.1
Increased Costs of Nonrenewable Energy	-488	18.6	10.6
Increased Penetration by Renewable Energy	-80	3.0	1.7
Range of Effects[c]			
Reduced Renewable Energy Costs	76 to 121	2.9- 4.6	1.7- 2.6
Increased Costs of Nonrenewable Energy	-452 to -520	17.2-20.0	9.8-11.4
Increased Penetration by Renewable Energy	-39 to -124	1.5- 4.7	0.9- 2.7

[a]Remember that these are cumulative changes in GNP. They are, thus, not strictly comparable to GNP results for any one year. However, this comparison was included to demonstrate the magnitude of cumulative GNP changes.

[b]Effect of the energy change specified but averaged across all combinations of initial energy conditions.

[c]Lowest and highest effects of the energy change across all combinations of initial energy conditions.

Increased costs for nonrenewable energy forms is clearly an instance of absorbing larger volumes of economic resources in order to provide each unit of energy services. Increased penetration by renewables is also an instance of a cost increase. Since each unit of energy services provided from renewable sources requires, on average, more input of economic resources than the conventional energy displaced, the use of more renewable energy involves a net claim on capital, labor and other economic resources.[4] Although higher costs for nonrenewables and increased penetration by renewables both involve a net economic cost, the far larger role of nonrenewables in the energy system means that the rise in nonrenewables costs has a substantially larger effect on the overall economy.

The same set of forces apply in determining the impact of reduced costs for renewables. Here, however, the change involves a cost reduction, or a net release of capital, labor and other resources. These resources become available for use in other forms of production, raising the output potential of the economy as a whole. Thus, a decrease in renewables costs will benefit the economy. This benefit will be larger, the larger the reduction in unit costs.

The second part of Table 5-6 shows the range of economic costs associated with each of the energy changes. In each case, the range of variation in the effects is small relative to the average level. For example, the average economic benefit from a reduction in renewables costs is $98 billion, but this benefit can range from $76 to $121 billion (1980$) depending on pre-existing energy conditions. Increased penetration by renewables has an average cost of $80 billion but this cost can vary anywhere from $39 to $124 billion (1980$) depending on other energy conditions. Finally, the economic cost of higher nonrenewables costs can range from $452 to $520 billion, with an average value of $488 billion (1980$), depending on the nature of accompanying energy conditions. In each of these cases, the range of variation around the average value is relatively small. Thus, it is the specific change in energy conditions that primarily determines the direction and magnitude of the economic impact.

---

[4]Although the cost disadvantage of renewable energy resources (on average) greatly diminishes (and, for some technologies, completely disappears) over time, on average over the 1980-2000 period, renewable energy is more expensive per unit delivered energy services than the nonrenewable energy it displaces.

The nature of the interaction among energy changes is shown in more detail in Table 5-7. The average effect of reduced costs for renewable energy is to increase cumulative GNP for the 1980 to 2000 period by $98 billion in 1980 dollars (see Table 5-6). However, this effect varies with the initial conditions into which the reduction in renewables costs is introduced. Consider first how the level of nonrenewables costs influences the economic impact. If nonrenewables costs are at Reference Case levels, (i.e., nonrenewables costs follow the Baseline Trend defined for this study), then the renewables cost reduction leads to an economic benefit of $105 billion, but if nonrenewables costs are at the higher level (e.g., the LHC Case) the renewables cost reduction leads to a smaller benefit, of $90 billion (1980$). Thus, reductions in renewables costs have a larger economic benefit for lower, rather than higher, cost levels for nonrenewable energy. Similarly, reductions in renewables costs have a larger economic benefit if higher, rather than lower, levels of penetration by renewable energy were to occur.

Although these conclusions (i.e., the economic benefits of lower costs for renewable energy are greater (1) when nonrenewable energy costs are lower and (2) when renewables market penetration is higher) may seem somewhat difficult to understand initially, they become clearer once one recalls certain causal relationships. First, a reduction in renewables costs, for a given level of use of renewables, releases economic resources for other productive uses. However, when higher prices for energy from nonrenewable sources prevail, general economic productivity is lower (than if nonrenewable energy costs are lower). Thus, the productivity of the released resources (released because of lower renewables costs) is lower (than if lower nonrenewable energy costs prevail). It follows, therefore, that the economic benefit of reduced renewables costs is lower, the higher the level of costs and prices for nonrenewable energy.

For the interaction between renewables costs and penetration, the causal relationships are more straightforward. A given reduction in unit costs of obtaining energy from renewable sources will release more economic resources, the larger the number of units of renewable energy. Thus, the economic benefit from reduced costs of renewables is larger, the greater the level of use or penetration of renewable energy.

The other sections of Table 5-7 have similar interpretations. Higher nonrenewable energy costs do more economic damage for lower, rather than higher, levels of renewable energy costs. The reason for this is that the volume of economic production is slightly larger under the lower renewables costs, so the productivity loss

Table 5-7

INTERACTIVE EFFECTS OF CHANGES IN ENERGY CONDITIONS
(Change in Cumulative Real GNP, 1980-2000,
in Billions of 1980 Dollars)

Effect of Reduced Costs for Renewable Energy			
	Nonrenewables Costs:	Baseline Trend	105
		High	90
	Renewables Penetration:	Lower	82
		Higher	114
Effect of Higher Costs of Nonrenewable Energy			
	Renewables Costs:	Most Likely	-481
		Low	-493
	Renewables Penetration:	Lower	-515
		Higher	-460
Effect of Increased Penetration by Renewable Energy			
	Renewables Costs:	Most Likely	-96
		Low	-66
	Nonrenewables Costs:	Baseline Trend	-106
		High	-53

Note: Figures show the average cumulative real GNP effect of the indicated energy change under the indicated initial conditions.

caused by higher nonrenewables prices affects a larger economy, leading to a larger loss in total output. (The difference in cumulative GNP loss as compared to the Case where both nonrenewable and renewable costs are higher, however, is very small.)

Higher costs for nonrenewables do less economic damage, the higher the level of penetration by renewables. There are two reasons for this result. The first is that increased use of renewables permits a reduction in the use of conventional energy so that the volume of economic resources absorbed by a given rise in the unit cost of nonrenewable energy is less than in the case of lower renewables penetration. The second reason is that a larger volume of use of renewables itself imposes an economic cost, so the total volume of economic activity is less. Therefore, the productivity of the additional resources absorbed when nonrenewable energy costs increase is lower under conditions of higher market penetration of renewables. It follows that total economic loss caused by the absorption of these additional resources is lower.

The economic costs imposed by higher renewable energy market penetration also vary with accompanying energy conditions. This cost is lower, the lower are renewables costs. Also, the cost is lower, the higher are nonrenewables costs. Higher penetration by renewables involves replacing lower cost conventional fuels with higher cost renewables. This causes an economic loss. But, if renewables costs are reduced, even though they remain higher than the costs of conventional energy, the amount of the economic loss is also reduced. Similarly, the cost differential between renewables and nonrenewables is less when nonrenewables costs are higher, so the economic damage done by increased penetration by renewables is smaller when nonrenewables costs are higher.

CONCLUSIONS

The major conclusions relating to economic performance that can be drawn from the information developed in Chapters 3, 4 and 5 are as follows:

- Economic growth continues under all of the energy conditions investigated for this research. Under the worst conditions (High Penetration assumptions for renewable energy and High Cost Trend assumptions for nonrenewable energy -- HPHC Case), the annual rate of increase in economic growth is only about 0.1 percentage points lower than for the most favorable conditions (LC Case). Thus, under the worst conditions the economy (as measured by real annual GNP) grows by about 73 percent rather than 76 percent (for the most favorable case) over the 1980 to 2000 period. This relatively small difference demonstrates the remarkable ability of the economy to successfully adjust to large changes in energy conditions. Thus, GNP effects tend to be smaller in relative terms than the changes in the energy conditions that bring about these economic effects.

- Differences in economic performance among the eight scenarios investigated here, although relatively small in relative terms, are quite large in absolute terms. Cumulative GNP losses for the least favorable (HPHC) Case as compared to the most favorable (LC) Case are nearly $700 billion (1980$).
- The economic effect of increased costs for nonrenewable energy is considerably larger than the economic effects of changes in the cost and/or quantity conditions for renewable energy resources. This occurs because, for all the cases investigated, nonrenewble energy continues to be the predominate (in excess of 80 percent) form of energy used in the nation.
- Reduced costs for renewable energy have a beneficial effect on economic performance.
- Increased utilization of renewable energy above the level specified in the Reference Case has an adverse effect on the economy.
- Increased costs for nonrenewable fuels and conversion technologies have an adverse economic effect.

6
Economically Efficient Policies toward Renewable Energy

The economic analyses in the preceding chapters examined the effects of changing the market penetration levels and the costs of renewable energy sources. These impact analyses were conducted for two different price trends for conventional (or nonrenewable) fuels. One of the principal reasons for performing these analyses was to obtain insights and draw conclusions about appropriate government policies toward renewable energy. This chapter investigates the policy conclusions relating to renewable energy resources. It should be noted that our perspective is economic. Therefore, policies are assessed in terms of their effects on real incomes and production in the economy, as measured by real GNP, over the period 1980 to 2000. Of course, other, noneconomic objectives and impacts (such as environmental impacts) will be relevant to any policy decisions concerning federal support for renewable energy, but these other objectives are not taken into account in this study.

INCREASED PENETRATION OF RENEWABLE ENERGY

"What is the benefit from increased penetration of renewable energy sources?" This is an important question for the design of national energy policy. The answer to it may determine how much we pay for energy in the future and from what sources we obtain it. The simulations developed for this study compared "Lower" and "Higher" renewable energy market penetration scenarios (10.8 quads in 2000 for the Lower Market Penetration Cases and 15.1 quads for the Higher Market Penetration Cases, an increase of 40 percent) under a range of costs for renewable and nonrenewable energy resources. Comparison of economic performance between the two penetration levels (or utilization levels) for renewables permits the payoff from increased renewable penetration to be assessed.

The macroeconomic impacts of increased utilization of renewable energy are given in Table 6-1 for four sets of conditions, each condition representing a specific combination of renewable energy cost and nonrenewable energy cost conditions. The last column gives the overall economic effect of raising the level of use of energy from renewable sources. Each impact is negative. Thus, over the entire range of renewable and nonrenewable cost conditions studied in this research, increasing the utilization rate (or market penetration) of renewable energy sources above the Lower Market Penetration levels specified in the Reference Case results in a reduction in aggregate production and real incomes in the economy. In short, in all circumstances, economic performance is adversely affected by greater use of renewables. There is some variation in the magnitude of the impact. Increased penetration does least economic harm when renewable costs are low and nonrenewable costs are high, while more harm occurs when renewable costs are relatively high (at "Most Likely" levels specified for this research) and nonrenewable costs are relatively low (at the levels specified for the Baseline Trend). In all cases, the economic penalty for using more renewable energy is sustainable in that the rate of economic growth is only slightly slowed. However, the cumulative costs (in terms of GNP foregone between 1980 and 2000) are significant in absolute magnitude, ranging up to $124 billion, in constant 1980 dollars.

The reasons for the adverse economic effect of increased renewables penetration have been discussed in previous chapters. The source of this adverse economic effect is straightforward. The additional economic resources required per million Btu of energy supplied by renewables are greater than the resources released from each million Btu of conventional fuel displaced. On balance, the provision of the same volume of energy services requires a larger quantity of economic resources under increased renewables penetration. This leaves less economic resources for the rest of the economy and, consequently, permits a smaller volume of finished goods and services to be produced. The restructuring of the energy system toward increased supplies from renewable sources, therefore, results in lower overall economic productivity and reduced production and real income potential.

Because an increase in the penetration of renewable energy is associated with reduced real GNP, this implies that any energy policy or policies aimed solely at inducing or forcing increased penetration by renewable energy sources is economically inefficient. Even policies that give tax or other public subsidies to

Table 6-1

PAYOFF FROM INCREASED RENEWABLES PENETRATION
(Change in Total Cumulative Real GNP, 1980 to 2000, in billions of 1980 dollars)

Energy Cost Assumptions			
Renewables Cost	Nonrenewables Cost	Impact Calculation	Effect of Increased Penetration
Most Likely	Baseline Trend	HP vs. R	-123.5
Most Likely	High	HPHC vs. HC	-67.6
Low	Baseline Trend	HPLC vs. LC	-90.7
Low	High	HPLHC vs. LHC	-39.4

Notes:

1. The four outcomes give the effect of increased penetration by renewables under each combination of conditions for renewable costs and nonrenewable energy prices.

2. Impact calculation refers to the method of finding the numerical impact of higher penetration, e.g., HP vs. R means total real GNP in HP Case minus total real GNP in the Reference Case.

3. The penetration effect column gives the cumulative change in total real GNP over the 1980 to 2000 period due to the increase in renewables penetration, other conditions remaining unchanged.

renewable energy technologies and fuels and so reduce the cost of renewables to the purchaser do not alter the total economic resource cost of renewables. Thus, such policies do not reduce the economic burden caused by increased penetration. The preferred policy position in these circumstances is to have no policy, i.e., not to attempt to increase the use of renewables. This is not to deny the possibility of other economic benefits from increased use of renewables. For example, they might cause a reduction in oil imports and, so, in world oil prices or they might reduce GNP losses in the event of another oil import disruption. But such benefits would have to be quantified and compared with the certain economic costs of increased use of renewables in developing an effective and efficient overall renewables policy. In the absence of information on any price or supply-disruption related economic benefits from increased use of renewables, we must conclude that it is economically inefficient to promote increased penetration by renewable energy sources.

REDUCTION IN COST OF RENEWABLES

Another category of policy conclusions that can be drawn from this research involves those measures aimed at reducing the total unit costs of securing energy from renewable sources. This is distinct from measures, such as tax subsidies, intended to increase the penetration of renewables using existing technologies by reducing costs only to actual users (through some form of subsidy). Cost reduction measures include approaches such as research and development to achieve technological and manufacturing improvements that reduce the average cost of energy provided from renewable sources. The economic effects of cost reduction policies are analyzed in this study by comparing the impacts of lowering renewable costs while keeping renewable penetration and nonrenewable costs unchanged. This analytical specification is designed to examine the overall economic benefits provided by a reduction in renewable costs alone. The actual specification of the cost reductions is given in Chapter 2. The capital cost reductions range from about 10 percent for solar thermal space heating equipment to more than 30 to 40 percent for the more exotic renewable technologies, such as ocean thermal electric conversion (OTEC). Most of the capital cost reductions are in the range of 10 to 25 percent.[1]

Table 6-2 presents information on the overall economic effects of reducing renewable energy costs for each of the four combinations of the energy system (i.e., four combinations of nonrenewables costs and renewables penetration) analyzed for this study. The impact of lower renewables costs is positive in all cases. Thus, no matter what the existing conditions concerning renewables penetration and nonrenewables costs, any reduction in renewables costs helps economic performance, permitting higher levels of production and real incomes. The overall cumulative benefits (from 1980 to 2000), as indicated by real GNP gains, range from \$77 to \$121 billion (1980\$). Within this range, the impacts fall into two categories: when renewables penetration is at the Reference Case level, the gain from lower costs for renewable technologies is between \$77 and \$88 billion (1980\$) while at the higher penetration level the gain from lower costs is correspondingly higher -- between \$105 and \$121 billion (1980\$). The economic gain from reducing renewable technology costs also varies according to whether the costs for nonrenewable fuels and conversion technologies are at their Reference Case or higher levels. The economic gain from reducing costs for renewables is greater by about \$15 billion (1980\$) when nonrenewable energy costs are at the levels specified for the Reference Case.

The reason for the economic benefit from lower renewable energy supply costs is straightforward -- lower costs mean that the same energy service output can be obtained from a smaller quantity of economic resources. Therefore, economic resources are released for use elsewhere, permitting a higher level of nonenergy production. On balance, total output from the economy rises even though the input of economic resources remains unchanged. In short, productivity, production and real incomes are increased. The reason for the greater benefit when renewables penetration is higher is simply that the same cost reductions per unit of renewable energy apply to more units of energy, thereby releasing more economic resources for use elsewhere. Eventually, any reduction in the cost of

---

[1]Although the Low Renewables Cost (LC) Case also includes a reduction in the rate of real inflation for biomass fuels (from Reference Case levels), virtually all of the economic effects for this Case are caused by the capital cost reductions for renewable energy technologies.

Table 6-2

PAYOFF FROM REDUCED COSTS FOR RENEWABLE ENERGY TECHNOLOGIES
(Change in Total Real GNP, 1980 to 2000,
in billions of 1980 dollars)

Energy Assumptions			
Renewables Penetration	Nonrenewables Cost	Impact Calculation	Effect of Cost Reduction
Lower	Baseline Trend	LC vs. R	88.0
Lower	High	LHC vs. HC	76.8
Higher	Baseline Trend	HPLC vs. HP	120.5
Higher	High	HPLHC vs. HPHC	105.1

Notes:

1. The four outcomes noted in the right-hand column summarize the effect of reductions in renewables costs under each combination of conditions for renewable penetrations and nonrenewable costs.

2. Impact calculation refers to the method of finding the numerical impact of reductions in renewables costs, e.g., LC vs. R means total real GNP in LC Case minus total real GNP in the Reference Case.

3. The cost reduction effect column gives the cumulative change in total real GNP over the 1980 to 2000 period due to the reduction in renewables costs, other conditions remaining unchanged.

providing necessary inputs to the economy yields an economic benefit which is reflected in higher levels of economic production, real incomes and real GNP.

These results demonstrate that any reduction in the cost of securing energy from renewable sources will have an economic benefit. This suggests a possible payoff from policy measures designed to lower the cost of renewable energy. However, this result in itself is not a useful policy guideline since the same reasoning can be applied to any product or resource -- the economy benefits from reductions in unit costs of supplying hamburgers or cassette recorders in just the same way as it benefits from lower costs for renewable energy. Thus, this type of economic benefit is not a sufficient condition to prescribe or justify specific government policy measures intended to reduce costs.

However, some valuable quantitative information concerning policy design can be obtained from these analytical results. This information concerns the question of how much we should, as a nation, be willing to pay to secure the indicated reductions in the cost of renewable energy supplies. Presumably, these cost reductions are not free (if they were, producers would adopt them anyway and there would be no apparent role for the government to play). Rather, cost reductions can be secured by research and development, by materials improvements and by improvements in manufacturing, fabrication, installation and maintenance techniques. Each of these means of cost reduction itself involves costs. The central issue for policy design, thus, concerns the financial outlay that is warranted in order to achieve these cost reductions. In other words, what is the maximum cost that it is economically efficient to incur in order to secure the indicated reductions in the cost of energy from renewable sources?

The benefit, in real GNP increase, from the renewable energy cost reductions we analyzed, is from $77 to $88 billion (1980$) (depending on the level of conventional energy costs), when renewable energy supplies about 10.8 quads of energy in the year 2000 (Reference Case level). If the overall economic benefit is taken to be $77 billion (1980$) then the maximum outlay that would be justified in securing the cost reductions would be this same amount ($77 billion). But, in addition, it must be recognized that outlays intended to secure cost reductions may or may not work. Each type of activity intended to bring about cost reductions has some probability, less than one, of being successful. The determination of an efficient set of policy expenditures must take account of both the magnitude of the outlay and the probability of success. Therefore, the total outlay, C, that would be economically justified in securing the cost reductions is given by:

$$C < \text{Prob} * 77$$

where "Prob" is the probability of the outlays, C, actually achieving the cost reductions. The cost, C, in billions of 1980 dollars, includes government outlays and private outlays. It includes costs summed over the entire planning horizon, which includes research and development costs, production-related investments and training incurred in the attempt to secure the cost reductions.[2]

Thus, both the probability of success and the benefits from success enter into the calculation of the maximum economically efficient outlay associated with policies intended to secure reductions in the costs of energy obtained from renewable sources. For an illustrative range of probabilties, the outlay ceiling in billions of 1980 dollars is:

Probability	Outlay Ceiling
0.05	$ 3.9
0.10	7.7
0.25	19.3
0.50	38.5

For a low probability of success, say 5 percent, total outlays of up to $3.9 billion are justified in the attempt to secure cost reductions. If the probability of success is higher, say 25 percent, then outlays of up to $19.3 billion would be justified. Or, if spending would be expected to provide a 50 percent probability of the cost reductions, then this spending would be justified if it were less than $38.5 billion.

This type of approach provides some useful broad guidelines for government policy toward renewable energy. Cost reductions in the supply of energy from renewable sources are economically desirable, but it must be recognized that there is a cost associated with achieving these cost reductions. Considerations of economic efficiency place a ceiling on the outlay that is justified in trying to achieve the cost reductions. The probability of success and the economic benefit of

---

[2] In reality, the choice would be more complicated since the time flow of costs and benefits should also be considered through some discounting mechanism.

success in reducing renewables costs jointly determine this ceiling. The information contained in this ceiling is potentially useful for making policy decisions about whether or not, or at what level, to fund policies designed to secure reductions in the costs of renewable energy resources.

A related application of this information concerns decisions about the allocation of public funds among different functional areas. Consider, for example, the case in which there is $1 billion of policy resources to be allocated. Spending these resources on attempts to reduce renewables costs might have a 5 percent probability of success. The expected benefit would therefore be 0.05 * 77 = $3.9 billion. There might also be other policy activities (energy-related or not) competing for these same resources, for example, a subsidy for energy conservation. This use of the resources might have an expected benefit of $2 billion. In this case, it would be economically preferable to devote the policy resources to the renewables area. But, if the expected benefit from conservation were $5 billion, then it would be preferable to direct the policy to conservation and not to renewables. These are just examples, but the principle is illustrated. The application of policy benefit information to the question of where best to spend scarce public funds and related public resources (e.g., personnel resources) must consider both the probability of success and the potential payoff from investing in other policy initiatives.

PROTECTING THE ECONOMY AGAINST HIGHER CONVENTIONAL ENERGY COSTS

It has been suggested in some quarters that reducing the comparatively high costs of renewable energy supply and increasing the penetration by renewables will afford at least some protection to the economy against the adverse consequences of future increases in the costs for conventional, nonrenewable energy sources. This section examines this question and the related policy issue of whether renewable energy should be promoted to gain protection against higher prices for conventional energy.

The starting point for this analysis is an estimate of the order of magnitude of the economic damage caused by higher prices for conventional fuels. The simulations presented in earlier chapters provide this information. One set of simulations examined the effects of more rapid growth of conventional, nonrenewable energy prices than in the Reference Case. By 2000, for this high nonrenewable energy cost case, the crude oil price is 35 percent higher than in the Reference Case, natural gas prices are 36 percent higher, coal and nuclear costs are 23 percent higher. Similar cost increases were introduced for the capital costs of large scale energy production and conversion facilities. As a result, the average per Btu cost of energy produced from conventional sources is about 20 percent higher than in the Reference Case. These cost increases slow the rate of growth for productivity, production, real incomes and real GNP, causing a significant and permanent economic cost. Pairwise comparison of the simulations provides estimates of the total real GNP effects of higher conventional energy costs. The total real GNP loss over the 1980 to 2000 period due to higher conventional energy costs was estimated at around $488 billions of 1980$ (see Table 5-6). This corresponds to a reduction of about 0.1 percentage points in the average annual rate of economic growth (from 2.85 to 2.78 percent per year) and to real GNP in 2000 being about 1.3 percent less than without the higher prices. While these costs and effects are not overwhelming, they are certainly very large. Any action, policy or change that reduces the costs of higher conventional energy prices is, correspondingly, worthy of serious consideration.

Table 6-3 presents information on the real GNP effect of higher prices for nonrenewable energy. As discussed previously, these effects are negative for each of the cost or market penetration assumptions involving renewable energy sources. However, the economic costs of higher nonrenewables costs do vary depending on the assumptions for renewables costs and market penetration levels. The adverse economic impact of higher nonrenewables prices is reduced as the extent of renewables penetration is increased and as the unit cost of energy from renewables is increased.

The benefits from renewables in terms of reducing the economic damage from higher prices for conventional fuels are shown in Table 6-4. Increased penetration of renewables does, in fact, provide a positive benefit -- with more renewable energy being used, the economy suffers less damage as a result of higher prices for oil and other conventional fuels (and higher costs for nonrenewable energy technologies). The reason for this is that the increased penetration of renewables displaces some conventional energy input, so that higher prices for conventional fuels involve a lower (than otherwise) increase in the economic resources required to secure these conventional fuels. Therefore, when more renewables are used, any increase in conventional energy costs absorbs a smaller (than otherwise) volume of economic resources and imposes a smaller burden on the economy. Higher costs for nonrenewable energy still impose a significant economic cost but this cost is

Table 6-3

IMPACTS OF HIGHER NONRENEWABLE ENERGY PRICES
(Change in Total Real GNP, 1980 to 2000, in billions of 1980 dollars)

Renewables Assumptions		Impact Calculation	Effect of Higher Prices
Renewables Penetration	Renewables Cost		
Lower	Most Likely	HC vs. R	-508.4
Lower	Low	LHC vs. LC	-519.6
Higher	Most Likely	HPHC vs. HP	-452.5
Higher	Low	HPLHC vs. HPLC	-468.3

Notes:

1. The four outcomes noted in the right-hand column summarize the effect of higher prices for nonrenewable energy under each combination of conditions for renewable energy penetration and costs.

2. Impact calculation refers to the method of finding the numerical impact of higher nonrenewable prices, e.g., HC vs. R means total real GNP in HC Case minus total real GNP in the Reference Case.

3. The column "Effect of Higher Prices" gives the cumulative change in total real GNP over the 1980 to 2000 period due to the increase in nonrenewable energy prices, other conditions remaining unchanged.

Table 6-4

EFFECT OF DIFFERENT ASSUMPTIONS CONCERNING RENEWABLE ENERGY ON THE ECONOMIC COST OF HIGHER NONRENEWABLES PRICES
(Change in Total Real GNP, 1980 to 2000, in billions of 1980 dollars)

Gain from Increased Renewables Penetration	
Reference Case Renewable Costs	55.9
Low Renewable Costs Case	51.3
Gain from Lower Renewables Costs	
Reference Case Renewable Penetration	-11.2
Higher Renewable Penetration Case	-15.8

Note: 1. This information is based on Table 6-3.

2. The gain refers to the reduction in the economic damage (i.e., an increase in real GNP) due to higher prices for nonrenewable energy. A negative gain indicates an increase in economic damage (i.e., a reduction in real GNP).

reduced by about $51 to $56 billion (1980$), if a greater quantity of energy is obtained from renewable sources. In other words, the increased utilization of renewable energy reduces by about 10 percent the economic damage caused by higher prices for energy from conventional, nonrenewable sources.

In contrast, lower costs for renewable energy have the effect of increasing the economic damage caused by higher costs for energy from nonrenewable sources. Under lower cost conditions for renewables, the economic damage caused by higher nonrenewables prices increased by about $11 to $16 billion (1980$), or by about 3 percent. This is a relatively small effect but it is significant in that it is positive. Thus, reducing costs (but with no change in the market penetration) of renewables works to increase rather than reduce the damage done by higher prices for energy from nonrenewable sources. The reason for this seemingly counter-intuitive result is that an initial reduction in renewables costs stimulates the economy, permitting greater output which leads to a greater demand for energy input from conventional sources. The use of more energy from relatively more expensive conventional sources means that nonrenewable cost increases now are applied to a larger base, which further reduces GNP. This incremental loss in GNP caused by greater use of nonrenewable energy more than offsets the increase in GNP caused by lower renewable energy costs.

This information can now be integrated to address the specific policy question, "Are policies directed toward reducing the costs of energy from renewable sources beneficial or justified in terms of protecting the economy against the effects of higher costs for nonrenewable energy?" The answer is no. Policies that lower renewables costs, without a change in the level of use of renewables, actually accentuate the economic damage caused by higher nonrenewable costs. By contrast, if renewables penetration is increased (for a given level of costs of renewable energy) the damage cased by higher costs for nonrenewable energy is reduced. But, this does not provide a justification for increased penetration. First, increasing penetration of renewables involves, in itself, a net cost of from $91 to $124 billion (1980$) (see Section 6.1, above). Securing a subsequent benefit of around $53 billion (1980$) from the reduction in damage done by cost increases for nonrenewables, still leaves a net cost of at least $38 billion. Second, increased market penetration by renewables involves a certain cost of from $91 to $124 billion, in order to secure a possible benefit of around $53 billion. The benefit is possible, not certain, because the rise in costs of nonrenewables (over the levels prevailing in the Reference Case) is itself only a possibility. If the probability that this rise in nonrenewables costs, over the levels prevailing in the Reference Case, actually occurs is p, then the expected net gain (in billions of 1980 dollars) from increased renewables penetration is:

$$\text{Expected gain} = p * \$53 - \$107$$

(where $107 is the approximate mid-point of the cost range of $91 to $124 billion). Since p cannot exceed one, this net gain is negative in all situations. Even for relatively high probabilities of the high nonrenewables price path occurring, the net result of higher market penetration for renewables is a substantial loss. For example, if p = 0.5 the net economic loss (or net negative gain) is $81 billion (1980$).

This analysis suggests that policies promoting renewable energy, either in the sense of cost reductions or of increased penetration, are not economically justifiable if their purpose is to provide protection against the economic damage done by increasing costs for nonrenewable energy. However, in this analysis we have not considered another possible benefit of increased supplies for renewable energy resources. This benefit would accrue if a reduction in oil imports occurred in response to policies aimed at increasing use of renewable energy. This reduction in oil imports by the U.S. (which has been the world's largest importer of oil) could lead to slower growth in world oil prices and/or to a lower probability of a forced disruption in imported oil supplies. However, such a linkage between increased supplies of renewables and either a lower growth path for oil prices or a lower probability of imported oil supply disruptions has yet to be proved. Thus, in terms of the cost and quantity conditions examined in this analysis, the conclusion remains that public policy designed to promote renewable energy supply as a protection or insurance against changes in conventional energy supply conditions is not <u>economically</u> justified.

EFFICIENT POLICY TOWARD RENEWABLE ENERGY

These investigations into economically efficient public policy toward renewable energy must be considered in determining the appropriate role and direction for government policy toward renewable energy. The analysis conducted for this research has led to some specific policy conclusions. Given the probable costs of renewable energy resources (even under relatively optimistic assumptions);

(1) It is not economically efficient public policy to promote increased use (market penetration) of renewable energy sources.
(2) There is a potential benefit from reductions in the cost of securing energy from renewable sources. Policy focussing on these cost reductions is justified only if the public and private outlay involved in securing the cost reductions is less than the probability of securing that cost reduction multiplied by the economic benefit from the cost reductions. If a policy measure satisfies this criterion, then there is economic justification for pursuing the policy.
(3) Renewable energy cannot be promoted as a means of protection against the economic damage caused by increased prices of conventional energy.
(4) It may be justifiable to promote renewable energy as a means of protection against the economic damage caused by a supply disruption (e.g., oil embargo) for imported energy sources but only if such a disruption has a high probability of occurrence and would cause significant economic damage and if increased utilization of renewables would significantly reduce energy imports (rather than mainly substitute for a domestic energy resource such as coal).

    Even so, using renewable energy resources as a substitute for imported energy may not be the most cost-effective means of providing protection to the economy in the event of a supply disruption. If the probability of a supply disruption is relatively high and if such a disruption would have significant adverse economic effects, then a systematic search should be conducted to determine the most cost-effective means of protection. For example, stockpiling energy (the idea behind the Strategic Petroleum Reserve) may be a much cheaper means of protection against supply disruption than investment in renewable energy capacity.

Given these results, what is the most appropriate course for government policies toward renewable energy? On economic grounds, there is never a case for policy measures which try to increase the penetration of renewable energy, but there may be a case for policy measures directed to reducing the costs of energy from renewable sources. Whether or not policies directed to renewable energy cost reduction actually are justified, in turn, depends on whether the public and private outlays required to achieve these cost reductions are less than the expected economic benefit of the cost reductions achieved. Only if this condition is met would it be economically efficient to undertake such a policy.

Appendix: Background Material and Sources for Chapter 2

Most of the information presented in Chapters 3 through 6 of this book is the result of original research by the authors. However, much of the background data for this study (presented in Chapter 2) has been developed on the basis of an extensive search through published and unpublished sources. The principal data sources are noted in the bibliography included as part of this Appendix. The sources in this bibliography represent only a part of the many references we reviewed, but they are indicative of the scope of our background research.

DEVELOPMENT OF TECHNOLOGY COST ASSUMPTIONS

The technology costs presented in Tables 2-5A and 2-5B were not based on original research (e.g., engineering analyses) conducted as part of this study. Rather, every effort has been made to exploit the work of other researchers. Thus, from published and unpublished sources we collected data on current and future operating, maintenance and capital costs and current and future operating efficiency for energy supply technologies. A selection of the most relevant of these sources is included as part of this Appendix.

Data from the different sources often was inconsistent. For example, cost estimates made several years ago did not always reflect the most recent information on inflation rates, technological progress or operating experience. In some cases, costs were expressed in current dollars (including an allowance for inflation) while in other cases costs were expressed in constant dollars. Furthermore, equipment included, operating efficiency, and/or capacity factors were not always similar.

The first purpose of our review of these data was to eliminate these inconsistencies. After the data was collected, the information from the various sources was adjusted to reflect common assumptions concerning

- equipment
- operating efficiency
- capacity factors
- scale or size
- interest rates
- inflation rates
- construction schedules
- operating environment
- year of startup for facility operation.

Chapter 2 (pp. 21 through 24) provides some of the broad guidelines which were used to make these adjustments. Further, once cost estimates from the various sources were adjusted to a comparable basis, medians, means, ranges and standard deviations for cost estimates were derived.

In the case of currently available technologies, relatively small variations in current costs were found, so that current "national average" costs could be specified with confidence within a narrow range for nearly every technology. However, for all technologies, wide variation was found in estimates of future costs. Obviously, defining costs for any technology twenty years hence cannot be accomplished with certainty. For this reason, estimates for future costs and other information were analyzed for each technology under consideration to evaluate both the range of uncertainty for future costs and, more important, the likely direction of this uncertainty. Thus, we were interested in whether it was more likely (at least from the perspective of a "consensus" forecast, where such a forecast could be derived) that costs would be lower or higher than currently foreseeable costs. Once the direction of this variation was established, the magnitude was defined on the basis of the evaluation of observable factors which would have a bearing on costs and/or informed opinion about future trends affecting costs.

An example of the type of analysis conducted for a specific set of technologies is set forth below.

Analysis of Agricultural/Industrial Process Heat (AIPH) Costs

There are a wide variety of solar AIPH technologies, depending on the temperature required and other process considerations. Some of these technologies are commercially available now, such as low temperature flat-plate collectors, similar to those which are used to provide domestic hot water and space heat to residences. For temperatures of 25° to 100° C, efficient flat plate collectors with multiple glazings and selective absorptive surfaces are adequate. For slightly higher temperatures -- 50° to 175° C -- evacuated tube collectors are suitable. For still higher temperatures one needs to use line focus type collector systems, such a parabolic troughs, fresnel lens, and multiple reflectors (heliostats). Under development are solar ponds for low cost, low temperature heat (although some researchers are suggesting use of this technology for generation of electricity using fluids with low boiling temperatures), as well as point focus, and central receiver systems for high temperatures up to 500° C and more.

In trying to define current costs for solar AIPH systems, one encounters the problems common to all solar active systems. Variations in estimates for current costs can be the result of a number of factors, including:

- technology -- differences in cost may result if systems use tracking dish collectors vs. parabolic troughs, for instance.
- design temperature -- low temperature collectors may have different costs than high temperature collectors.
- size -- there may be economies of scale for larger installations, since fixed costs (e.g., storage facilities) can be spread over a larger installation.
- location -- there will be significant differences in performance at different locations. Development of national cost numbers is especially difficult, given the fact that many cost estimates are developed only for specific locations, usually in the Southwest.
- process -- the process for which the solar AIPH equipment is being used to generate heat may require specialized equipment. This will cause significant variations in costs.
- type of backup equipment -- specialized equipment may be required for different backup equipment.
- storage -- the amount of storage may vary from system to system. Generally storage costs are an increasing function of storage time above about two to four hours of storage. Also, the type of storage equipment (e.g., batteries vs. heat sinks) may cause variations in cost.

- purpose -- if the solar AIPH equipment is used for preheat only in combination with conventionally fired equipment, costs would differ than if the solar AIPH were used in a similar situation for complete heating of process fluids.

In addition to these factors, future projected cost estimates and cost goals are often based on technological improvements or technological breakthroughs as compared with systems that are currently being built. This factor further complicates cost comparisons for future years.

Representative cost estimates were collected from more than two dozen sources for solar AIPH. The cost estimates were all of a similar order of magnitude. We have adopted the federal 1981 solar costs goals (as specified by the U.S. Department of Energy and the Solar Energy Research Institute) as our standard for solar AIPH costs, with the following qualifications:

- in all cases the higher end of the range for solar costs was used to indicate the federal cost goal (representing a site with the characteristics of Ft. Worth, Texas).
- for 1990 and 2000 a ten percent increase over the solar cost goal was used to represent significant market penetration of solar AIPH at less favorable sites.

For our optimistic solar cost scenarios, a ten percent decrease below federal solar cost goals was used to represent increased penetration at more favorable sites (in the South and Southwestern United States). In other words, this scenario assumes that a very large proportion of total market penetration will occur in the Sun Belt.

The resulting costs are:

	1980[1]	1985	1990	2000
Base Case	200-250	180	110	80-90
Optimistic	200-250	180	100	65-70

These costs allow for increased utilization of shallow solar ponds for process heat. Design lifetimes are

---

[1]Market penetration is not expected to be significant before 1985 for solar AIPH.

assumed to be 20 years. O&M costs are assumed at 3 percent of initial capital costs for the Base Case and 2.5 percent in the Optimistic Case.

Similar analyses were performed for all major technologies in the following three groups:

- renewable (centralized and decentralized) technologies
- synthetic fuel conversion technologies
- electric generation technologies.

In all cases we analyzed both currently available technologies and technologies which were likely to become commercially available prior to the year 2000. The sources reviewed included, for every major technology, information from each technology's vendors and potential vendors, advocates and detractors, as well as information from more neutral sources. Current costs were first established (where appropriate) and future cost estimates were developed to be consistent with current costs and with trends in manufacturing techniques, materials cost trends, labor input cost trends, forecasts for technological progress, and other factors. Where appropriate, costs were specified for mass-produced equipment. Thus, although certain home-built decentralized equipment (such as solar collectors) has been fabricated by individuals at lower costs than for commercially available equipment, such cost estimates were not given great weight in specifying costs for this research.

DEVELOPMENT OF ENERGY SUPPLY/DEMAND SCENARIOS

Some of the major sources for the energy supply/demand scenarios developed for this research are noted in Chapter 2. As noted throughout the text, the scenarios analyzed in this book represent only a few of the many feasible (in a technological or engineering sense) supply/demand combinations which could evolve over the next two decades in the United States. The scenarios we have developed are achievable though evolutionary shifts by energy consumers and providers.

Although these scenarios are certainly not representative of the extremes in energy supply/demand relationships posited by some observers (i.e., they do not incorporate "crash" programs to develop synthetic fuels, or a complete transition to any particular set of renewable or nonrenewable energy sources), the scenarios we have developed have several important attributes. First, they are consistent with "consensus" energy forecasts developed by the U.S. Department of Energy and other sources for planning purposes over the long term. Second, our scenarios incorporate movement from more vulnerable energy sources (e.g., imported petroleum products) to more secure souces of supply. Third, our scenarios incorporate significant differences in the energy supply mix by the year 2000. In particular, all scenarios incorporate much greater use of new technology energy sources.

Again, as with the specification of technology costs, an extensive literature search was conducted to develop these scenarios. Some of these sources are included in the bibiliography which follows.

SELECTED REFERENCES FOR CHAPTER 2

Adelman, M. H., et al, "Energy Self-Sufficiency, An Economic Viewpoint," Technology Review, May 1974, 76(6) 22.

Adler, T.J., et al. The Direct and Indirect Costs of Transporting Wood Chips to Supply a Wood-Fired Power Plant. Research Program on Technology and Public Policy, Thayer School of Engineering, Dartmouth College, July 1978.

Allen, E. L. Energy and Economic Growth in the United States. MIT Press, 1979.

Allen, Edward L., et al. U.S. Energy and Economic Growth: 1975-2010. Institute for Energy Analysis, Oak Ridge Associated Universities, September 1976.

American Gas Association, Comparison of Initial Capital Investment Requirments for New Domestic Energy Supplies: 1980 Update. 1980.

American Gas Association, Planning and Analysis Group, "A Comparison of Capital Investment Requirements for Alternative Domestic Energy Supplies." May 5, 1978 (unpublished).

Anderson, Carl J., et al. An Assessment of U.S. Energy Options for Project Independence. Prepared by Lawrence Livermore Laboratory (University of California) for the Energy Research and Development Administration, May 1975.

Anderson, T. D., et al. An Exploratory Study of Cost Targets for Solar Electric Power Plants, Summary Report. Prepared by Oak Ridge National Laboratory, March 1977.

Associated Universities, Inc. Reference Energy Systems and Resource Data for Use in Assessment of Energy Technologies. NTIS #PB221 422 (also AET-8), May 1972.

Auer, P.L., A.S. Manne, and O.S. Yu. Nuclear Power, Coal, and Energy Conservation. Electric Power Research Institute, SR-34, March 1976.

Balcomb, J.D., and R.D. McFarland, "A Simple Empirical Method for Estimating the Performance of a Passive Solar Heated Building of the Thermal Storage Wall Type." Proceedings of the Second National Passive Solar Conference, Philadelphia, PA, March 1978.

Banker's Trust Company, "U.S. Energy and Capital Forecast: 1978-1982." 1977.

Banker's Trust Company, "U.S. Energy and Capital: A Forecast, 1980-1990." 1979.

Battelle-Columbus Laboratories, Systems Study of Fuels from Sugarcane, Sweet Sorghum, Sugar Beets and Corn. Volume II, March 31, 1977.

Bayraktar, et al (eds)., Energy Policy Planning. New York, Plenum Press, 1979.

Bechtel Corporation, Coal-Fired Power Plant Capital Cost Estimates. Electric Power Research Institute, AF-342, January 1977.

Bechtel Corporation, Resource Requirements, Impacts, and Potential Constraints Associated with Various Energy Futures. Prepared for the U.S. Department of Energy, August 1978.

Bentley, Geoffrey K. New England Rural Energy Sources. New England Congressional Caucus, February 1979.

Bhagat, N., et al. Evaluation of Technical Data in the DFI and PIES Models. Prepared for the Energy Information Administration, U.S. Department of Energy, by Brookhaven National Laboratory, 1979.

Bland, L.T., and R.K. Riley, "Future U.S. Demand Patterns and Use of Hydrogen." Paper presented at the Conference on the Hydrogen Economy, Miami, FL, March 1974.

Bliss, C., and D.O. Blake. Silvicultural Biomass Farms, Volume V, Conversion Processes and Costs. Prepared by The MITRE Corporation for the U.S. Department of Energy, May 1977.

Boyer, V.S, "Electric Power Generation Economics." Distributed by the Atomic Industrial Forum.

Bradley, Darby, and David Stevens. Wood for Energy, An Interim Report on the Whole-Tree Harvesting Experiment. South Duxbury, VT, March 1, 1978.

Breyer, A. V. Electric Power Generation Economics. Philadelphia Electric Energy Information and Education Service, Philadelphia, PA, 1978.

Bryant, Alden, "Investment and Economy in Energy Production: The Solar Thermal Electric Option." Paper presented at the International Solar Energy Society Conference, Denver, CO, August 28-31, 1978.

Buchanan, Deborah L. *A Review of the Economics of Selected Passive and Hybrid Systems*. Prepared by the Solar Energy Research Institute for the U.S. Department of Energy, January 1979.

Burwell, C.C., "Solar Biomass Energy: An Overview of U.S. Potential," *Science*. Vol. 199, March 10, 1978.

Cameron Engineers, Inc., *Shale Oil Status Report*. Denver: Rocky Mountain Oil and Gas Association, March 1978.

Cameron Engineers, Inc., *Synthetic Fuels, Report by the Subcommittee on Synthetic Fuels of the Committee of the Budget, U.S. Senate*. September 27, 1979.

CIA, "The International Energy Structure: Outlook to 1985." 1979.

Cliff, W.D. *Effect of Generalized Wind Characteristics on Annual Power Estimates from Wind Turbine Generators*. Prepared by Battelle Pacific Northwest Laboratory for U.S. Department of Energy.

Committee on Nuclear and Alternative Energy Systems (CONAES), NAS, *Energy in Transition: 1985-2010*. W.H. Freeman, San Francisco. 1979.

Construction Manpower Demand System, *Forecasts of Cost, Duration and Manual Man-Hour Requirements for Construction of Electric Generating Plants: 1977-1981*. A project jointly sponsored by the Tennessee Valley Authority and the U.S. Department of Labor, January 1978.

Council on Environmental Quality, *Environmental Quality -- Ninth Annual Report*. Washington DC, 1978.

Council on Environmental Quality, "A National Energy Conservation Program - The Half and Half Plan." March 1974.

Council on Environmental Quality. *Solar Energy: Progress and Promise*. April 1978.

Curto, P.A., and Grant Miller, "An Update of OTEC Baseline Design Costs." METREK Division, The MITRE Corporation, McLean, VA, 1978.

Dantzig, G.B. Stanford Pilot Energy/Economic Model. Prepared by Stanford University for Electric Power Research Institute, May 1978.

Davy Power Gas, Inc., "Winkler Technology for Clean Fuels from Coal," in Synthetic Fuels Processing. 1975.

Deonigi, D.E., et al. Capital Requirements and Capital Formation for Various Assumed U.S. Energy Supply Patterns. A Battelle Energy Program Report, June 1976.

Duane, J.W., and M.A. Karnitz, "Domestic Gas Resources and Future Projection Rates," Power Engineering. January 1975, 79(1) 36-39.

Duchesneau, Thomas D. Competition in the U.S. Energy Industry. Ballinger, 1975.

Dupree, W.G., and J.A. West, "U.S. Energy Puzzle." Proceedings of the Division of Refining - 1973, Philadelphia, PA. American Petroleum Institute, Washington, DC, May 1973.

Edison Electric Institute, "Economic Growth in the Future." 1975.

Edison Electric Institute, "Economic Growth in the Future - II." June 1980.

Electric Power Research Institute, Combined Cycle Power Plant Capital Cost Estimates. Palo Alto, CA, January 1977.

Electric Power Research Institute, Economic Modeling of Electricity Production from Hot Dry Rock Geothermal Reservoirs: Methodology and Analysis. Palo Alto, CA, September 1979.

Electric Power Research Institute, Perspectives on Utility Central Station Photovoltaic Applications. Palo Alto, CA, 1978.

Electric Power Research Institute, Supply 77. EPRI EA-634-SR, May 1978.

Electric Power Research Institute, Technical Assessment Guide. 1978.

Exxon Corporation, "Energy Outlook: 1975-1990." Houston 1975.

Exxon Corporation, "World Energy Outlook." December 1979.

Federal Power Commission, Hydroelectric Power Resources of the United States, Developed and Undeveloped. Washington, DC, November 1976.

Flaim, Silvio, et al. Economic Feasibility and Market Readiness of Solar Technologies, Volume 1. Solar Energy Research Institute, Golden, CO, September 1978.

Fluor Engineers and Constructors, Inc. Economic Studies of Coal Gasification Combined Cycle Systems for Electric Power Generation. Electric Power Research Institute, AF-642, January 1978.

Fluor Engineers and Constructors, Inc. Economics of Current and Advanced Gasification for Fuel Gas Production. Electric Power Research Institute, AF-244, July 1976.

Ford Foundation, Nuclear Power Issues and Choices. Nuclear Energy Policy Study Group, Washington, DC, 1977.

Ford Foundation, A Time to Choose, America's Energy Future. Ballinger Publishing Co., Cambridge, MA, 1974.

Foreman, K.M., and B.L. Gilbert, "Technical Development of the Diffuser Augmented Wind Turbine (DAWT) Concept." Paper presented at Wind Energy Innovative Systems Conference, Colorado Springs, CO, May 1979.

Garate, John A. Wind Energy Mission Analysis. Prepared by the General Electric Co., Space Division, for the U.S. Energy Research and Development Administration, February 1977.

General Electric Co. Requirements Assessment of Wind Power Plants in Electric Utility Systems. Prepared for the Electric Power Research Institute, January 1979.

General Electric Co., Space Division. Applied Research on Energy Storage and Conversion for Photovoltaic and Wind Energy Systems, Final Report, Volume III, Wind Energy Systems with Energy Storage. Prepared for National Science Foundation, January 1978.

Gibbs and Hill, Inc. Economic Comparison of Coal and Nuclear Electric Power Generation. January 1980.

Goodman, Alvin S., M. Llewellyn Thatcher, and Ruben S. Brown. Small Hydropower Projects in New York State. ASCE Convention, April 2-6, 1979, Preprint 3549.

Gordian Associates. Overcoming Institutional Barriers to Solid Waste Utilization as an Energy Source: Executive Summary. Prepared for the U.S. Department of Energy, November 1977.

Gritton, E. C., R. Y. Pei, et al. A Quantitative Evaluation of Closed Cycle Ocean Thermal Energy Conversion (OTEC) Technology in Central Station Applications. Rand Corporation, Santa Monica, CA, May 1980.

Hartley, Fred L., "Shale Oil -- A Synthetic Fuel Whose Time Has Come." Speech to The National Academy of Engineering, Washington, DC, November 2, 1978.

Hittman Associates, Inc., Environmental Impacts, Efficiency, and Cost of Energy Supply and End Use. November 1974.

Holt/Procon. Energy Conversion and Economics for Geothermal Power Generation at Heber, California; Valles Caldera, New Mexico; and Raft River, Idaho -- Case Studies. Prepared for the Electric Power Research Institute, November 1976.

ICF, Inc., "Oil Import Reduction: An Analysis of Production and Conservation Alternatives," in Synthetic Fuels, Report by the Subcommittee on Synthetic Fuels of the Committee on the Budget, U.S. Senate. August 1979.

Johnson, L.R., et al. Energy Supply and Environmental Impacts, Unconventional Sources. Prepared by the University of Idaho for the Northwest Energy Policy Project, 1977.

Joint Committee on Atomic Energy, Understanding the National Energy Dilemma. Joint Committee Print, August 1973.

Keeny, S. M., et al, Nuclear Power Issues and Choices, Ballinger, Cambridge, MA. 1977.

Lancaster, Angela, "The Economics of Future Electricity Generation from Nuclear and Coal Power Plants." Congressional Research Service, December 30, 1976.

Landsberg, H.H., et al. Energy: The Next Twenty Years. Ford Foundation, Ballinger, 1979.

Lindsay, Franklin A., "Financing High-Cost, High-Risk Energy Development," Harvard Business Review. November-December 1978, pp. 161-170.

Lovins, Amory B. Soft Energy Paths: Toward a Durable Peace. Ballinger Publishing Co., Cambridge. 1977.

Macrakis, M.C., ed., Energy: Demand, Conservation and Institutional Problems. MIT Press, 1979.

Maine Wood Fuel Corporation, Electric Power Generation From Forest Products and Cellulosic Materials. December 1978.

Massachusetts Institute of Technology, Ocean Thermal Energy Conversion: A State-of-the-Art Study. For Electric Power Research Institute, Palo Alto, CA, June 1979.

Massachusetts Technology Development, A Guide to Wood Fuel Conversion in Moderate Size Facilities: Some Considerations. Prepared for the New England Regional Commisison, January 1979.

McSweeney, J.J. Costs of Nuclear vs. Coal-Fired Power Plants in the Mid-1980s, for Colliers Management Corporation, January 1977.

Mead, W. J., and A. E. Utton (eds.), U.S. Energy Policy: Errors of the Past, Proposals for the Future. Ballinger, Cambridge, MA. 1979.

Medville, D., et al. Comparative Economic Assessment of Ethanol from Biomass. Prepared by The MITRE Corporation for U.S. Department of Energy, September 1978.

MidWest Research Institute, Continuation of Systems Study of Fuels From Grasses and Grains: Phase 1, Phase 2 and Phase 3, Draft Final Report. Prepared for Solar Energy Research Institute, Department of Energy, January 31, 1979.

Miller, Saunders. The Economics of Nuclear and Coal Power. New York: Praeger Publishers, 1976.

The MITRE Corporation, Solar Energy: A Comparative Analysis to the Year 2020. Prepared for the U.S. Department of Energy, March 1978.

The MITRE Corporation, Systems Descriptions and Engineering Costs for Solar Related Technologies, Volume IV: Conventional Technologies for Electric Utility Applications. Prepared for the Division of Solar Energy, U.S. Energy Research and Development Administration, June 1977.

The MITRE Corporation, Systems Descriptions and Engineering Costs for Solar Related Technologies, Volume IX: Biomass Fuels Production and Conversion. Prepared for the Division of Solar Energy, U.S. Energy Research and Development Administration, June 1977.

The MITRE Corporation; CONSAD Research Corp.; International Research and Technology, Inc., Environmental Data for Energy Technology Policy Analysis. Prepared for U.S. Department of Energy, 1978.

Moody, J.D., and R.F. Geyer, "Petroleum Resources. How Much Oil and Where?" Technology Review. March-April 1977, 77(5) pp. 38-45.

Mooz, William E. Cost Analysis of Light Water Reactor Power Plants. U.S. Department of Energy, June 1978.

NASA/ASEE - Auburn University, "MEGASTAR, The Meaning of Energy Growth - An Assessment of Systems, Technologies, Requirements." NASA/CR-120338, September 1974.

NASA/ASEE - Auburn University, "Terrestial Applications of Solar Technology and Research (TERRASTER)." NASA/CR-129012, September 1973.

National Academy of Engineering, Synthetic Liquid Fuels from Coal. November 2, 1978.

National Petroleum Council, Guide to NPC Report on U.S. Energy Outlook. December 1972.

New England Energy Congress, Final Report to the New England Energy Congress. 1979.

Nutter, J.F., and C.S. Waitman. Oil Shale Economics Update. Prepared by TOSCO Corporation for American Institute of Chemical Engineers, Southern California Section, April 18, 1978.

Oak Ridge National Laboratory and United Engineers and Construction, Power Plant Capital Costs Trends and Their Sensitivity to Economic Parameters. Prepared for the Division of Reactor Research and Development, U.S. Atomic Energy Commission, October 1974.

Office of Technology Assessment, U.S. Congress, Application of Solar Technology to Today's Energy Needs. Government Printing Office, June 1978.

Office of Technology Assessment, U.S. Congress, An Assessment of Oil Shale Technologies. June 1980.

Pelley, William E., et al. U.S. Energy and Capital: A Forecast: 1978-1982. Bankers Trust Co., 1978.

Perl, Lewis J., "The Economic Viability of Nuclear Energy." Paper presented before a seminar for members of Congress and their staffs sponsored by the Scientists' Institute for Public Information, June 7, 1976.

Pleatsikas, C.J., et al. A Study of Capital Requirements of Solar Energy. Urban Systems Research & Engineering, Inc., prepared for U.S. Department of Energy, July 1979.

Pleatsikas, C.J., D. Funkhouser, et al. Estimates of the Costs of Renewable Energy Technologies for New York State. Urban Systems Research & Engineering, Inc., prepared for New York State Energy Research and Development Authority, 1980.

Prince, B.E., et al. Survey of Nuclear Fuel Cycle Economics: 1970-1985. Prepared by the Oak Ridge National Laboratory, March 1977.

Proceedings of the Fifth Ocean Thermal Energy Conversion Conference, 1978.

Reichle, Leonard, "The Economics of Nuclear Power." Presented to the Conference on the Outloook for Nuclear Power, December 14, 1976.

Resources for the Future, Review of FEA National Energy Outlook 1976. March 1977.

Ross, P.N., Westinghouse Electric Corporation, "The Nuclear Electric Economy." Paper presented at the Conference on the Hydrogen Economy, Miami, FL, March 1974.

Rossin, A.D., and T.A. Rieck, "Economics of Nuclear Power," Journal of the American Association for the Advancement of Science. Vol. 201, August 18, 1978.

Roy, D.L., "The Nation's Energy Future." U.S. Atomic Energy Commission (USAEC), WASH-1281, December 1973.

Rudasill, C.L., Coal and Nuclear Generating Costs. Electric Power Research Institute, PS-455-SR, April 1977.

Sant, Roger, "The Least Cost Energy Strategy: Minimizing Consumer Costs through Competition." Mellon Institute, 1979.

Sargent and Lundy, Inc. Trends in Electric Generating Costs. February 1979.

Schurr, S.H., et al, Energy in America's Future, the Choices Before Us. Resources for the Future, Inc., Johns Hopkins University Press, 1979.

Science and Public Policy Program, University of Oklahoma, Energy Alternatives: A Comparative Analysis. Prepared for various federal agencies, 1975.

Shell Oil Company, "National Energy Outlook - 1978."

Shurmeier, H.M., Jet Propulsion Lab. Testimony delivered to the Energy Development and Applications Subcommittee of the Committee on Science and Technology, U.S. House of Representatives, March 1, 1979.

Solar Energy Research Institute, Photovoltaic Venture Analysis. SERI/TR-52-040, Vols, 1-3, July 1978.

Stanford Research Institute, Solar Energy in America's Future. Prepared for the Division of Solar Energy, Energy Research and Development Administration, March 1977.

Stanford Research Institute, Western Regional Energy Development. Vol. II, December 1975.

Stone and Webster. Economics of Nuclear Power. For the Atomic Industrial Forum, February 1979.

Stoughbaugh, R., and D. Yergin, eds. Energy Future. Random House, 1979.

Tabors, Richard D. Photovoltaic Power Systems: Review of Current Market Studies: Methodology for Long-Term Demand Projection. MIT Energy Laboratory Working Paper, May 1978.

Taylor, R.J., and L.L. Green. Problems of Hydroelectric Development at Existing Dams. Applied Physics Laboratory, Johns Hopkins University, Interim Report, November 1978, and Final Report, April 1979.

Taylor, Vincent. The Easy Path Energy Plan. Union of Concerned Scientists, 1979.

Teller, E., "Energy - A Plan for Action." Report to the Energy Panel of the Commission on Critical Choices for Americans, 1975.

Temple, Barker and Sloane, Inc., Study of the Electric Utility Industry: Demand, Costs, and Rates for the U.S. Federal Energy Administration. Prepared for the Federal Energy Administration, July 1976.

TRW, Inc., Systems and Energy Group, Cost and Performance Evaluation of Passive Solar Systems. Prepared for the Energy and Environmental Systems Division, Argonne National Laboratory, November 1978.

TRW, Inc., Systems and Energy Group, Utilization of U.S. Geothermal Resources. Electric Power Research Institute, EPRI ER-382, December 1976.

Tsou, P., et al, in 13th IEEE Photovoltaic Specialists Conference. IEEE, Washington, DC, 1978, pp. 1196ff.

United Engineers and Constructors, Capital Cost Addendum: Multi-Unit Coal and Nuclear Stations. Nuclear Regulatory Commission, NUREG-0245, June 1977.

United Engineers and Constructors, Capital Cost: Boiling Water Reactor Plant. Nuclear Regulatory Commission, NUREG-0242, June 1977.

United Engineers and Constructors, Capital Cost: High- and Low-Sulfur Coal Plants -- 1200 MWe. Nuclear Regulatory Commission, NUREG-0243, June 1977.

United Engineers and Constructors, Capital Cost: Low- and High-Sulfur Coal Plants -- 800 MWe. Nuclear Regulatory Commission, NUREG-0244, June 1977.

U.S. Atomic Energy Commission, LMFBR Program: Final Environmental Impact Statement, Vol. 4." Washington, DC, December 1974.

U.S. Atomic Energy Commission, Office of Planning and Analysis, "Nuclear Power Growth 1974-2000." WASH-1139(74), February 1974.

U.S. Department of Commerce, "Analysis of Past and Expected Future Trends in U.S. Energy Consumption, 1977-2000." February 1977.

U.S. Department of Commerce, Survey of Current Business. Washington, DC.

U.S. Department of Energy, Domestic Policy Review. 1978.

U.S. Department of Energy, "Energy Supply and Demand in the Short Term: 1979 and 1980." June 1979.

U.S. Department of Energy, Hydroelectric Plant Construction Cost and Annual Production Expenses in 1976. May 1978.

U.S. Department of Energy, Inexhaustible Energy Resouces Planning. 1978.

U.S. Department of Energy, Low Energy Future for the U.S. June 1980.

U.S. Department of Energy, Monthly Energy Review. Washington, DC.

U.S. Department of Energy, National Energy Plan II. 1978.

U.S. Department of Energy, "Short Term Energy Outlook." February 1980.

U.S. Department of Energy, Steam Electric Plant Construction Cost and Annual Production Expenses, 1976: 29th Annual Supplement. August 1978.

U.S. Department of Energy, Energy Information Administration, Annual Report to Congress. 1979, 1980.

U.S. Department of Energy, Energy Information Administration, Turbine Plant Construction Cost and Annual Production Expenses -- 1978. Washington, DC, October 1980.

U.S. Department of the Interior. Energy Perspectives 2. June 1976.

U.S. Energy Research and Development Administration, Comparing New Technologies for the Electric Utilities. December 9, 1976.

U.S. Energy Research and Development Administration, Office of Planning and Analysis, "Total Energy, Electric Energy and Nuclear Power Projections." February 1975.

U.S. Environmental Protection Agency, Alternative Futures and Environmental Quality. Office of Research and Development, Washington, DC, November 1973.

U.S. Federal Energy Administration, 1976: National Energy Outlook. February 1976.

U.S. Federal Energy Administration, Project Independence (PIES Model). November 1974.

Urban Systems Research & Engineering, Inc., Macroeconomic Impacts of Variations in the Cost and Market Penetration of Solar Technologies. Prepared for U.S. Department of Energy, July 1981.

Verplanck, W.K., and W.W. Wayne, Jr. Report on Turbogenerating Equipment for Low-Head Hydroelectric Developments. Stone and Webster Engineering Corp., Boston, MA, April 1978, under U.S. Department of Energy Contract EG-77-C-07-1692.

Wayne, W.W., Jr. Available Low-Head Hydro Technology. Stone and Webster Engineering Corp., Boston, MA.

Whittle, C.E., and D.B. Reister, "The IEA Energy Simulation Model: A Framework for Long-Range U.S. Energy Analysis." Institute for Energy Analysis, January 1975.

Williams, R bert H., "Industrial Cogeneration," from Annual Review of Energy, 1978. Vol. 3, pp. 313-356.

Wilson, C. L., et al, Energy: Global Prospects 1985-2000, McGraw-Hill, New York. 1977.

World Coal Study, Coal -- Bridge to the Future. MIT Press, May 1980.

Worthington, J. Dean, Public Service Gas and Electric of California, "Geothermal Development." Proceedings: EPRI Annual Geothermal Program Project Review and Workshops. Palo Alto: Electric Power Research Institute, January 1978.

About the Book and Authors

Solar Energy and the U.S. Economy

Christopher J. Pleatsikas, Edward A. Hudson, and Richard J. Goettle IV

If the United States is to be weaned from its dependence on foreign oil and insulated from the vagaries of OPEC price hikes, it must now begin to look not just at what energy technologies are available to replace fossil fuels, but also at what kinds of investment are required to make alternative energy sources a reality.

This book investigates the economic effects of investing in solar energy technologies, given a variety of assumptions about the cost of the technologies, the cost of fossil fuels (and electric generating capacity), and the amount of solar energy that can be supplied. Applying their economic model, the authors make long-range projections of economic growth, inflation, productivity, and capital investment in the 1980-2000 period. Their comparisons of economic growth in situations of differing combinations of energy cost and energy supply identify the potential economic effects of large solar energy investments that must be considered in the formulation of future energy plans for the United States.

Christopher J. Pleatsikas is an economist at Price Waterhouse and Company, working in the areas of energy and regulatory economics. Edward A. Hudson and Richard J. Goettle IV are both economic consultants specializing in economic forecasting, policy analysis and strategic planning. In addition, both have written extensively on topics concerning energy economics.

For Product Safety Concerns and Information please contact our EU representative GPSR@taylorandfrancis.com Taylor & Francis Verlag GmbH, Kaufingerstraße 24, 80331 München, Germany

Batch number: 10726011

Printed by Printforce, the Netherlands